# THE CALL OF
# EVEREST

As many have before and since, a climbing team ascends into clouds as they climb **Mount Everest.**

# THE CALL OF EVEREST

## THE HISTORY, SCIENCE, AND FUTURE OF THE WORLD'S TALLEST PEAK

### CONRAD ANKER
Foreword by Thomas Hornbein

NATIONAL
GEOGRAPHIC

WASHINGTON, D.C.

**MOUNT EVEREST**
8,850 m   29,035 ft

*South Summit*
8,750 m
28,707 ft

Dashed section of route
located behind ridge

*South Col*
7,906 m
25,938 ft

Camp IV
7,906 m
25,938 ft

*Northeast Ridge*

*Changtse*
7,583 m
24,879 ft

*West Ridge*

Camp II
Advance Base Camp
6,492 m
21,300 ft

*Khumbutse*
6,665 m
21,867 ft

*Rongbuk
Glacier*

*Western*

*Lho La*

Camp I
5,944 m
19,500 ft

*Lingtren*
6,749 m
22,142 ft

*Pumori*
7,165 m
23,507 ft

*Khumbu Glacier*

*Khumbu
Icefall*

Base Camp
5,364 m
17,600 ft

*Khumbu Glacier*

—— 1996 *Everest* Expedition, South Col Route

# The Route Up Everest

The first American expedition put 6 of its nearly 1,000 men on the 29,035-foot pinnacle. One pair labored up by the unconquered West Ridge and came down the other side—the first traverse of a major Himalayan peak. Others took the South Col route pioneered by Hillary and Tenzing.

When the lowland porters left, Sherpas and American mountaineers leapfrogged supplies up the forbidding slopes through a series of camps placed a day's climb apart. At Camp VI, the last before the summit, Sherpas struggled up to 27,450 feet with tents, oxygen, food, butane stores, and sleeping gear that made possible the final assaults. Routes up the South Col and West Ridge diverged at Camp II, the advance base.

To paint this panorama, Austrian artist Heinrich C. Berann made a detailed study of the area from a chartered airplane.

In 2012 a team sponsored by The North Face and the National Geographic Society endeavored a Legacy Climb in honor of this expedition's 50th anniversary. Members of the team summited via the South Col, but conditions made the West Ridge ascent too hazardous to try.

Lhotse
8,501 m
27,890 ft

Geneva
Spur

Nuptse
7,861 m
25,791 ft

Camp III
7,315 m
24,000 ft

C w m

Baruntse
7,168 m
23,517 ft

Cho Polu
6,695 m
21,965 ft

*Imja Glacier*

Namche Bazar
29 kilometers
18 miles

BERANN

Success! Tenzing Norgay holds out his ice ax into the thin air on the summit of Mount Everest on May 29, 1953, as he and Edmund Hillary became the first people to stand at the summit.

# CONTENTS

# EVEREST CALLS

### THOMAS HORNBEIN

n the spring of 2012 the National Geographic Society and The North Face sponsored the Legacy Expedition to Everest. Its goal and that of a companion expedition sponsored by Eddie Bauer was to revisit the 1963 American Mount Everest Expedition's journey nearly half a century before, with teams ascending from both the South Col route and the West Ridge. The plan, like ours in 1963, was to meet on the summit, with the West Ridge climbers descending, as did Willi Unsoeld and I, by the South Col route. Though the South Col part of this plan succeeded during one of the two brief windows of tolerable weather, excessive rockfall and poor snow conditions on the West Ridge made it too hazardous. This year's was not an uncommon experience; the success rate on this route over the past half century has been around 10 percent of about 60 attempts, and the chances of dying have proved to be about the same as those of reaching the summit.

This book, *The Call of Everest,* also celebrates that same 50th anniversary of the first American expedition to climb Everest (and also the 60th anniversary of the first ascent of Everest by Sir Edmund Hillary and Tenzing Norgay). This volume is a natural for the National Geographic Society, which was a major supporter of the 1963 American Mount Everest Expedition (AMEE), not least because of the participation of one of their own, Barry Bishop, who was on the team both as climber and photographer. His photo (opposite) of two tiny figures, Willi Unsoeld and me, on the crest of the West Shoulder, dwarfed by the mountain soaring seductively above captures for me the essence of what our West Ridge adventure was about—savoring uncertainty.

The AMEE was successful almost beyond our wildest dreams. On May 1 Jim Whittaker, along with Sherpa Nawang Gombu, became the first American to summit Everest. Three weeks later four more followed. Lute Jerstad and Barry Bishop topped out by the South Col route midafternoon on May 22. A tad tardy, at 6:15 p.m. Willi and I completed the first ascent of the West Ridge (via what Willi liked to refer to as Hornbein's avalanche trap, now known as the Hornbein Couloir). We headed down the South Col route, catching up with Lute and Barry, and we four finished it all off sitting out the night together above 28,000 feet in an unintended bivouac. We were very lucky to have survived the night and lucky that

**A SCOUTING PARTY** during the 1963 American Mount Everest Expedition appears as small figures in the foreground of Everest's forbidding West Ridge (opposite). The group inspects part of a new route that will be used by Willi Unsoeld and Tom Hornbein to reach the summit.

**THOMAS HORNBEIN STANDS** on the summit in 1963 (below). He would write, "The question of why we had come was not now to be answered . . . The answer lay not on the summit of Everest, nor in the sky above it, but in the world in which we belonged and must now return."

TOM HORNBEIN (WITH oxygen canister) shows the Sherpa members of the 1963 American expedition how to operate their supplemental oxygen equipment. While the use of supplemental oxygen in climbing Everest would be an object of debate, it placed the summit within reach for most climbers.

Willi and I had been able to pull off the first traverse of a major Himalayan peak.

Our ascent of a new route on Everest was a predictable stage in the evolution of humankind's relationship with mountains. The underlying theme of this quest is to perpetuate uncertainty: First we identify a goal worthy of our dreams, then figure out how to get to its base, and finally seek a way to its top. Once the mountain has been "conquered" (an abominable term), the next challenge is to find new paths to follow. This stage also begins to encompass style and, on Everest in particular, climbing without the aid of supplemental oxygen. Later in this evolution new uncertainties are added, like skiing or parachuting from the summit to be down in time for tea. Inevitably, as the human-mountain relationship reaches maturity, the advent of guided climbing brings less experienced aspirants to enjoy a taste of ultimate adventure.

Though none of this evolution is unique to Everest, what is unique is the explosion in numbers seeking to be guided to the top of the highest point on Earth. Everest is being loved to death. But it's not the mountain that's dying.

I must admit that until the events of the 2012 season, and in spite of the tragedy of 1996 and some subsequent years, I was able to rationalize the growing numbers and pragmatically accept that guided climbing was as inevitable on Everest as on every other attractive mountain in the world. I reckoned that less experienced but fit people could be guided up Everest with a reasonable margin of safety. I was impressed that the mortality rate did not go up despite the increased numbers of novices. I attributed this relative success to the strategies adopted by thoughtful guide services working together, for example fixing ropes over any terrain where a fall could carry you off.

I also imagined that the motivation of these current-day guided clients may not be all that different from my own. I had observed long ago that the intensity of this drive to succeed seemed to bear no relationship to the reasons driving us, whether to seek fame and fortune, to enjoy the intimidating beauty of one of Earth's wild places, or simply to engage in some inner spiritual quest where one seeks an answer to the question "Can I?"

One of the starkest contrasts between 1963 and now was captured by Mark Jenkins's description of events in 2012 (Chapter 7), and by photos, one of a thin, dark line: hundreds of climbers clipped to a fixed rope, snaking their way up the Lhotse Face and another showing climbers packed like sardines awaiting their turn to ascend the Hillary Step. Perhaps the biggest risk to climbing Everest these days is overcrowding and the risk of waiting in a queue, sometimes for hours. What fills me with even more dread is imagining that moment when these numbers high on the mountain intersect with the quirky forces of nature—a rogue storm, a massive avalanche—that are part of the roulette of climbing on big mountains. As with other natural disasters, perhaps the question is not whether but when.

As a flatlander growing up in the Midwest, I met mountains when I was 13 and fell in love. That love has defined my life, not just as an amateur climber of mountains but also in many other ways—in my profession as physician, researcher, and educator and with my family and many others who have added seasoning and helped me to grow old with, I hope, a modicum of grace. When I was a teenager, Everest was part of my fantasy world, not a place I figured I'd ever see, much less climb upon. Looking back from this 50th anniversary, I feel blessed to have been born at the right time and given the opportunity to share in such an adventure at a time when the mountain belonged just to us.

That Everest is past. Everest now, with its multitude, the social atmosphere at Base Camp and the ambience up high, is the antithesis of what mountains mean to me. My affair with this vertical world has been mostly a quiet one, wandering in wild places with like-minded companions. Even so, I think I understand the attraction this tallest of peaks has for those willing to brave its challenges; we share a similar fire to pursue dreams. This volume is aptly named—Everest "calls." So turn the page, and let Conrad Anker get you started.

> **"TO PUSH ONESELF TO WITHIN A WISP OF DEATH AND RETURN TO THE VALLEY IN THE WORLD BELOW IS TO SEE LIFE IN ITS RAW IMMEDIACY."**
> —SIR EDMUND HILLARY

# CHAPTER 1

# THE MEANING

Conrad Anker

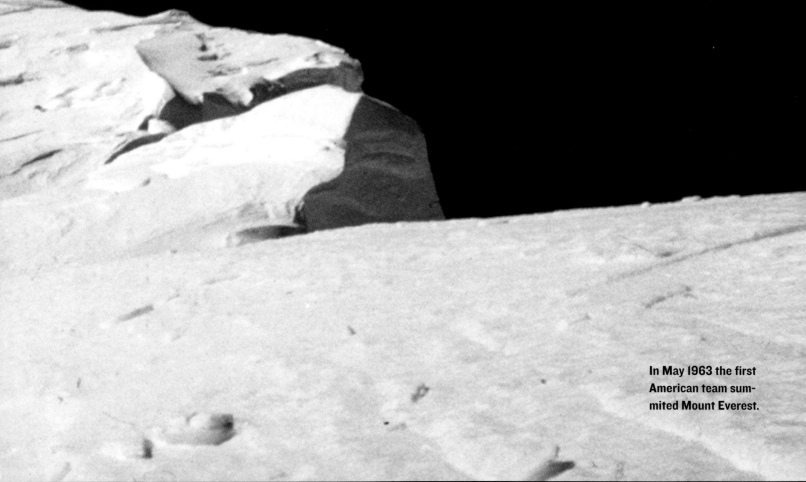

# OF EVEREST

In May 1963 the first American team summited Mount Everest.

ON MAY 26, 2012, I LOOK DOWN FROM THE SUMMIT OF MOUNT EVEREST TO THREE GLACIERS THAT HAVE SCULPTED THE MOUNTAIN. FOR THE PAST NINE AND A HALF HOURS I HAVE BEEN CLIMBING THE SOUTHEAST RIDGE IN NEAR-PERFECT WEATHER.

On May 26, 2012, I look down from the summit of Mount Everest to three glaciers that have sculpted the mountain. For the past nine and a half hours I have been climbing the Southeast Ridge in near-perfect weather. At an elevation of 8,850 meters, there is no higher place on our planet. The world literally drops away below. To the east, the robust Kangshung Glacier pushes moraine into Tibet and in the process creates small glacial lakes. To the north, the Rongbuk Glacier is solid in appearance, yet I know that it is moving, ever so slowly. To the south and west, the Khumbu Glacier, cascading down the southern flank of the Himalaya, provides sustenance to the people of Nepal and India. The frozen snow on which I stand may eventually join the Ganges, slowly making its way to the Bay of Bengal and the Indian Ocean. Perhaps this water will be recycled and deposited once again in the Himalaya to begin the timeless cycle of regeneration.

Standing on the apex of our planet is humbling. I'm starved of oxygen, depleted of reserves, unable to eat, and bound by anxiety. This is a dangerous place. Yet the symbolism of standing on top of the world gives me a chance to experience time on a cosmic scale. During the half hour I spend on the summit, I reflect on the mountain—how it came to be, its significance to humanity, and my

USING A FIXED rope for safety, mountaineer Conrad Anker descends an ice step in the Khumbu Icefall. Considered one of the most dangerous parts of the route to and from Everest's summit, the icefall is located about a third of the way down the Khumbu Glacier on the Nepali slope of Everest.

MEMBERS OF THE 1924 British Everest expedition gather at Base Camp in optimistic times, before their trip turned fatal. On June 8, 1924, Andrew "Sandy" Irvine (back row, far left) and George Mallory (standing next to Irvine) disappeared into the clouds about 800 yards from Everest's summit. Mallory's remains were discovered by Conrad Anker 75 years later.

personal connection to Everest. For the third time I have the opportunity to stand at this unique spot on the planet.

Humans frame time within the span of our own existence and, to a lesser extent, the history of humanity. We anthropomorphize time, as if what happens to humans is the only relevant measure. We are exhorted to live in the here and now. Yet on the upper reaches of the highest mountain, we live on borrowed time. Dillydally too long and we will die. When we face adverse situations, time is immediate. This immediacy provides a prism through which we can view our planet. How do we fit into the grand scheme of life? Everest, with its timeless immensity, highlights how insignificant human existence is. Standing on the summit, looking up through the troposphere to the blue, purple infinity of space, on a mountain of rock millions of years old, thrust up into the sky by a thin crust of earth floating on a moving mantle and carved away by gravity, I contemplate my place in the universe.

I feel insignificant. The mountains seem to have conquered us long before we set foot on them, and they will remain long after our brief existence. This indomitable force of the mountains gives us humans a blank canvas on which to paint the drive of discovery and, in the process, test the limits of human performance.

## THE IRRESISTIBLE CALL

Chomolungma, as the denizens who live within the shadow of Everest have referred to it for millennia, has always been revered as a spiritual abode. The mountain gained recognition within the Western mind only in 1849, when its measurements were announced by the Great Trigonometrical Survey of India. More than a century after it was quantified as the tallest point on Earth, Tenzing Norgay and Ed Hillary climbed Everest on May 29, 1953. After eight expeditions that cost 13 human lives, the goal had been reached at last—Earth's third pole. This landmark of exploration signified the culmination of terrestrial exploration. We had crossed the seven seas centuries earlier, mapped the ranges and rivers in due time, and reached the axes of our planet at the turn of the 20th century. Now the last and ultimate landmark had been reached.

# "IT WAS NOT GLORY WE SOUGHT, UNLESS IT BE THE COMMON GLORY OF MAN'S TRIUMPH OVER NATURE— AND OVER HIS OWN LIMITATIONS."

—BRIG. SIR JOHN HUNT, LEADER OF THE BRITISH EVEREST EXPEDITION, 1953

**FILM SHOT TWO** days after Mallory and Irvine left for the summit shows dejected members of the British team walking away from two sleeping bags formed into a signal to those below that there was no hope of finding their comrades.

The first ascent of Everest came at a time when humanity needed relief from two world wars. It was a unifying and inspiring event, signifying the drive to reach our greatest potential. Tenzing Norgay, a Sherpa born in the shadow of the great mountain, reached the summit with Edmund Hillary, an enthusiastic beekeeper and mountaineer from New Zealand. That an Easterner and Westerner teamed up for this historic ascent signified a new era of cooperation and a transition away from colonial rule. Times were changing, and Everest was symbolic of this change.

## AN AMERICAN DREAM

By the time the United States launched an expedition in 1963, nine people had reached the summit. On May 1, 1963, when Jim Whittaker, the first American, reached the summit of Everest, I was all of 156 days old. I wasn't aware of my own existence, and crawling about my crib was the extent of my climbing experience. But mountains were already part of my heritage.

Our family valued the serenity, beauty, and physical challenge of the mountains where my ancestors had settled, in Big Oak Flat, California, and we

THE 1963 AMERICAN team, sponsored by the National Geographic Society, ascends Everest to Base Camp. The expedition was spearheaded by Swiss climber Norman Dyhrenfurth and included 19 Americans, 32 Sherpas, and 909 porters carrying 27 tons (25 metric tons) of gear. The team's purpose was not only to reach the summit but to carry out scientific research in physiology, psychology, glaciology, and meteorology.

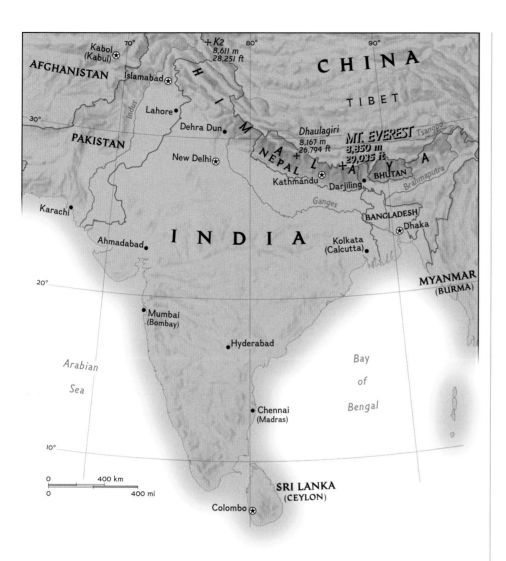

Map labels:
- Kabol (Kabul) — AFGHANISTAN
- 70°
- Islamabad
- K2 8,611 m 28,251 ft
- 80°
- CHINA
- TIBET
- 90°
- Lahore
- HIMALAYA
- Dehra Dun
- 30°
- Dhaulagiri 8,167 m 26,794 ft
- MT. EVEREST 8,850 m 29,035 ft
- Tsangpo
- PAKISTAN
- New Delhi
- NEPAL
- HIMALAYA
- BHUTAN
- Kathmandu
- Darjiling
- Brahmaputra
- Ganges
- Karachi
- BANGLADESH
- Dhaka
- Ahmadabad
- INDIA
- Kolkata (Calcutta)
- 20°
- MYANMAR (BURMA)
- Mumbai (Bombay)
- Hyderabad
- Bay
- Arabian
- of
- Sea
- Bengal
- Chennai (Madras)
- 10°
- 0    400 km
- 0    400 mi
- SRI LANKA (CEYLON)
- Colombo
- Indus

**THE HIGHEST PEAK** in the Himalaya, Mount Everest straddles the border of Nepal and Tibet. It got its name from Sir George Everest, a British surveyor general of India, although for centuries Tibetans have called it Chomolungma, often translated as "goddess mother of the world."

embarked on annual pack trips to camp in the High Sierra each summer. We climbed peaks together, and I remember asking my father as we looked out over the landscape, "Where will we go next?" I was wondering what the extension of our summer vacations in the gentle, granite-ringed meadows might be. Dad spoke of the Himalaya, the most dramatic mountain range on Earth, and pointed me to a book that was part of his library. With a seed of possibility planted, I would leaf through Tom Hornbein's illustrated book *The West Ridge* and dream of giant glaciated peaks and the remote people who lived beneath them. My eyes were opened to a world beyond North America. I dreamed of being out on the edge, and becoming a climber was the way to get there.

## WHAT'S BEYOND?

A scant 16 years after Tenzing and Ed's 1953 ascent, humans reached the moon. In July 1969 our family huddled around a flickering black-and-white television to watch it happen. Neil Armstrong and Buzz Aldrin walked on the moon. Yes,

exploration was alive and well. In a span shorter than a generation, humans had gone from the top of our planet to the one celestial body that Earth casts its shadow upon, providing proof that we exist within a greater scope. The next morning, as my mother recalls, I rode my bicycle around with added vigor as I pretended to be an astronaut. Lunar exploration had supplanted terrestrial exploration.

Still, Everest continued to offer a test of human limitations. In 1978, when Reinhold Messner and Peter Habeler climbed to the summit without the use of supplemental oxygen, unaided by life's most essential element, humanity was again inspired by their accomplishment. In 1980, when Messner returned to climb it alone, again without added oxygen, he pushed the standard a notch higher. A decade later Everest transitioned from the cutting edge of alpine climbing to a mountain that, given the right amount of training and resources, became a goal attainable by the average person. The early expeditions were an extension of impe-rial power, starting with the three pioneering expeditions of 1921, 1922, and 1924. As large national expeditions gave way to commercial guided climbs in the early 1990s, Everest became a mountain for the individual rather than the nation-state.

## THE MALLORY MYSTERY

In 1999, I was invited to take part in the Mallory and Irvine research expedition to the north side of Everest. Seventy-five years earlier, on the last of the initial three English Everest expeditions, George Mallory and his young partner, Sandy Irvine, climbed within striking distance of the summit when they disappeared into the clouds. They failed to return. Could they have made it, on June 8, 1924, 29 years before Hillary and Norgay? The question became one of the most debated in mountaineering history. We will never know whether they made the summit. The mystery remains, symbolic of Mallory's pithy answer offered when asked why humans choose to climb Everest. His four words—"Because it is there"—probe into the fundamental question of what human existence means.

My team set out to solve the mystery by finding a missing camera that may have documented a summit. Instead, on May 1, 1999, I discovered the frozen and preserved body of George Leigh Mallory at an elevation of 26,700 feet (8,140 meters) on the north side of Everest. In the half hour that I spent alone with his body, I felt connected to the expedition that long ago had set the stage for where I was that day. The drive to reach a summit and the planning that goes

> # "GETTING TO THE TOP IS OPTIONAL, BUT GET-TING DOWN IS MANDATORY."
>
> —ED VIESTURS, FIRST AMERICAN TO CLIMB THE WORLD'S 14 8,000-METER PEAKS WITHOUT SUPPLEMENTAL OXYGEN

## THEN & NOW    LADDERS

| 1963 | 2012 |
|------|------|

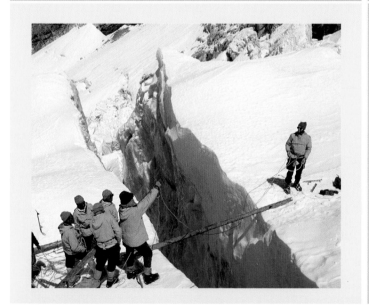 

In the 1963 American expedition led by Norman Dyhrenfurth, precarious logs served for the mountain's death-defying chasm crossings; in 2012, National Geographic's climbing team employed lightweight aluminum ladders, allowing for easy transport and safer passage. The ladders also served during tricky vertical ascents.

## "I LOVE THE MOUNTAINS BECAUSE THEY REMIND ME THAT I'M PART OF SOMETHING MUCH BIGGER THAN MYSELF."

–JULIAN COOPER, BRITISH PAINTER AND CLIMBER

into one goal—a goal that offers only intrinsic rewards in the face of great risk—were the same motivations that drew me to the mountains. Though separated by 75 years, Mallory and I were partners in our life's ambition.

### MEETING MALLORY

Frozen to the slope, arms outstretched and legs crossed, Mallory died doing what he loved. His family paid the price for his ambition and obsession. The proximity to death, the slow dance one has with mortality at high altitudes, is painfully familiar to mountaineers. I was humbled by the moment and felt deep respect for the man on whose shoulders the current generation of climbers stood. But we still did not know if Mallory had summited Everest, for we did not find the camera.

Continuing the quest, on May 17, 1999, I set out to climb the Second Step without the aid of an aluminum ladder that had been installed by the 1975 Chinese expedition. The Second Step, a formidable 90-foot cliff at 28,230 feet on the Northeast Ridge, was the most challenging barrier to the summit. As far as we knew, no one had climbed it without using the ladder. Had Mallory and Irvine been able to reach the top in 1924, they would have had to overcome this challenge. If we could climb it under the same conditions that Mallory and Irvine had encountered, we hoped to unlock the mystery. I led the pitch and gave it a fair shot, but at the last moment I stepped on the ladder despite my effort not to.

## HIGHS AND LOWS

I reached the summit in the late afternoon with my partner, Dave Hahn. A patch of snow forms the top of the world, and barely a thousand people had ever been there. Our celebration was brief. The weather was worsening, Dave was ill, and there were no other climbers on the route. We began our descent, feeling alone and desperately close to getting cut by the edge I had set out to find.

In October of the same year, I survived an avalanche on another peak in the Himalaya, the Tibetan giant Shisha Pangma. But it took the lives of my two climbing partners, Alex Lowe and David Bridges. Alex had been my closest friend, and once again I saw the edge of existence. This time I felt the suffering that death brought to family and friends, and my own life changed dramatically. In the spring of 2001, I married Alex's widow, Jennifer, and adopted his three young children, Max, Sam, and Isaac. I continued to climb but with a renewed sense of caution, and the welfare of others took a greater role in my own sense of purpose. With Jenni and the Alex Lowe Charitable Foundation that she created, I was able to give back to the mountain culture to which I felt indebted. In 2003, we launched the Khumbu Climbing Center to teach technical climbing skills and mountain safety to the Sherpas and indigenous people of the Himalaya. As I traveled to Nepal, winter after winter, to share my skills and love for climbing, the people of the Himalaya began to feel like my family.

In 2007, I returned to the north side of Everest to film a biographical documentary on Mallory and his life. I had a second go at the Second Step. We removed the ladder, returning the cliff to the state that Mallory and Irvine would have encountered in 1924. With great effort, I was able to free-climb the overhanging crack without using the ladder, in conditions

ANKER (OPPOSITE) IN 2007 during his second summit of Everest. On this expedition, he successfully climbed the Second Step, a 90-foot cliff on the Northeast Ridge. No one else had climbed the cliff without the aid of a ladder, but Anker knew that if Mallory made it to the summit, that's how he would have had to do it. Anker wanted to see if it was possible—and he proved that, although admittedly challenging, it was.

AS THE NUMBER of people attempting to summit Everest continues to increase, so does the number of rescues. After dropping off injured climbers at Base Camp, a helicopter (below) turns around to retrieve more climbers farther up the mountain. The year 1996 was the deadliest on the mountain to date: Eight people died in a two-day period when a storm struck, causing whiteout conditions and plummeting temperatures.

TRAILS OF HEADLAMPS coursing through tents at night (above) show how bustling Base Camp can be. While seen by many as the jump-off point for their long ascent to the peak, many people take the trek to 17,598 feet to see the Khumbu Icefall, in between Everest's West Shoulder and Mount Nuptse. Anker (opposite) has Everest in his sights.

similar to Mallory's attempt. It was difficult and challenging in the rarefied air, yet aided by the benefit of modern equipment and technique, the feat was still far easier than what Mallory would have faced. Our team reached the summit on June 14, giving me a second chance to stand at the apex of our planet.

## CHANGING MEANING

When I was born in 1962, the world population was 3.1 billion people. The population of Nepal was 10 million and a grand total of 9 people had stood on the summit of Everest. By the time I reached the summit in 2012 on my third expedition, the world population had crested 7 billion and the population of Nepal had swelled to more than 26 million. A total of 106 people summited on that same day—May 26, 2012—bringing the total number of summits to more than 6,000 in 50 years' time. Popular sentiment, prodded by the press and inevitable deaths that still occur on the mountain, held that with so many climbers, Everest was no longer relevant. It had gone from an epic, daring adventure to an easily attainable trophy.

Everest will always be a focal point of human attention because it is the highest point on our planet, and as such it retains tremendous drawing power. While the cognoscenti in climbing circles no longer respect its ascent, it still has an amazing allure for the nonclimber. Knowing the intrigue and allure that Everest still has with citizens of our planet, I set out to create a Legacy Expedition to honor the first American Mount Everest expedition of 1963.

## THE LEGACY CLIMB

Our 2012 team brought together several educational components. The earth sciences of Everest, the composition of the rocks and the forces that drove them to the apex of the planet, became the focus of Dave Lageson and the team from Montana State University. By collecting a rock profile that ranged from the summit to the base, he sought a fuller understanding of the rock and how it got to its position. Sharing this information with the National Science Foundation and creating a curriculum for fourth- and fifth-grade students provided an exciting way to engage in science.

Bruce Johnson and the Mayo Clinic team studied the effects of high altitude on human physiology, especially how the cardiopulmonary system reacts to such duress. Studying how the healthy body reacts to altitude allows researchers to better predict and monitor disease. Our team included rock climbers with no previous altitude experience. Would they be able to withstand the stark demands? The tests done on the climbers tracked the same adaptations and reactions that people with heart disease face. Testing medical devices in the demanding environment of Everest would make them about as durable as it gets.

Combining high-altitude physiology, earth sciences, and the adventure of climbing in the real-time world of the Internet provided the team with the chance to reach millions of people. Fifty years ago the story was told via newspapers, magazines, and books. Communication in 2012, for better or worse, was immediate. A nearby cell tower now allows Sherpas and climbers to keep in touch with their families on a daily basis. Satellite phones, about the size of a stick of butter, provide instant communication from the upper reaches of the mountain. We shared the Base Camp experience with thousands of people via a real-time Google chat. And on May 26, 2012, I met up with my friend Dave Hahn on the summit. It was his 14th ascent of the mountain and my 3rd. Under blue skies, we were able to spend a bit more summit time together than we had 13 years earlier.

# "TO STRUGGLE AND TO UNDERSTAND —NEVER THIS LAST WITHOUT THE OTHER..."

—GEORGE MALLORY

After this, I realized that my connection with Everest had come to a turning point. I was as old as the 1963 Everest expedition. I had reached the summit without supplemental oxygen, an accomplishment that required more effort and personal sacrifice than my first two climbs. In my mind, I had treated the mountain with greater respect, climbing the peak by fair means.

My connection to the Khumbu region of Nepal, my bonds to the Sherpas who inhabit it, and the friendships stemming from two decades of climbing there have left me with meaningful and lasting impressions. While I have traveled to the Himalaya to climb ice, rock, and mountains, it is the interactions with indigenous populations that I have valued most. In 2012 I joined the advanced Sherpa team to ascend the Lhotse Face. We climbed together, fixing rope for all of the subsequent climbers. I experienced firsthand the work these climbers do to prepare the route each season. From Tenzing Norgay Sherpa in 1953 to Panuru Sherpa, our sirdar (or expedition leader) in 2012, Sherpa climbers have been integral to Everest history. Their hard work and dedication provide the foundation for all successful ascents. Every climber owes a debt to these unsung heroes of Everest.

## THEN & NOW  OXYGEN SYSTEM

### 1953

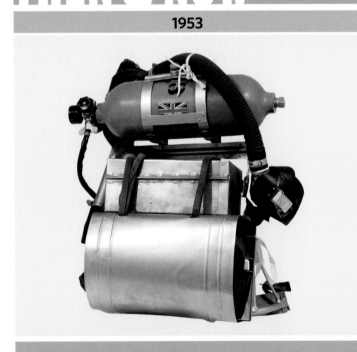

### 2012

In the British Everest expedition of 1953, a 35-pound closed-circuit oxygen system added hefty weight to the climbers' gear. Fifty-nine years later, the National Geographic team ascending the same peak benefited from Summit Oxygen's lightest system in the world—a slim aluminum canister weighing 5.5 to 7.6 pounds depending on its three- or four-liter capacity.

# FOREVER EVEREST

Despite the naysayers, Everest still offers challenges for the aspiring alpinist. Fantasy Ridge, originating from the Kangshung Glacier in Tibet and connecting to the Northeast Ridge, is perhaps the largest unclimbed challenge in the Himalaya. Repeating routes in better style—without the use of fixed ropes, high-altitude support, and supplemental oxygen—still offers climbers unlimited opportunities to test their mettle.

Everest remains a symbol of exploration and discovery. Its increase in popularity has reflected the growth of humanity on our planet. The challenges of resource use and overcrowding on Everest mirror what civilization below is facing. The Himalaya, a low-latitude, high-altitude mountain range, is at the forefront of anthropogenic climate change. The glaciers are subject to global warming and a changing atmosphere. These challenges are the next "Everest" for human discovery and exploration. The same drive that brought humans to the summit will motivate future generations on to further unknowns. If our expedition to Everest has helped to keep the flame of inquisitiveness and motivation alive, we will have contributed to the collective human drive for a better life here on Earth.

Please join us as we explore Everest in the coming pages.

**WITH THE KHUMBU** Icefall behind him, a member of the 2012 North Face/National Geographic expedition heads into the Western Cwm, a broad, flat glacial valley that is traversed by climbers heading up the southeast route of the mountain. Despite its icy appearance, sunny, windless days can get hot in the valley, fatiguing climbers before the push for the summit.

# CHAPTER 2

# THE BIRTH

David R. Lageson

# OF EVEREST

The roof of the world as seen from a jet flying overhead

ANCIENT PERSIAN GEOGRAPHERS CALLED THE MOUNTAINS OF CENTRAL ASIA BAM-I DUNYA— THE ROOF OF THE WORLD. EXTENDING FOR NEARLY 1,800 MILES, THE KARAKORAM-HIMALAYAN-TIBETAN BELT IS A DYNAMIC GEOLOGICAL LABORATORY, FORMED BY THE CLOSURE OF TWO ANCIENT OCEAN BASINS AND THE COLLISION OF NUMEROUS BLOCKS OF EARTH'S CRUST.

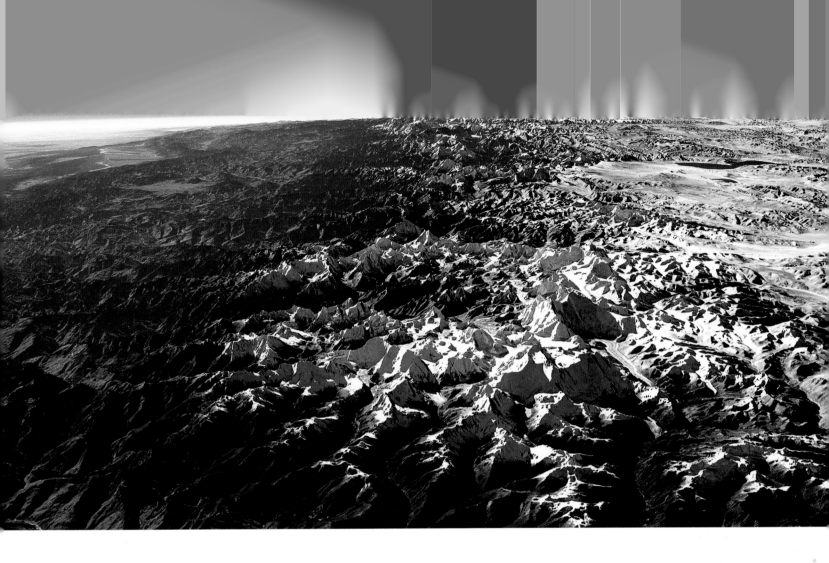

Ancient Persian geographers called the mountains of central Asia Bam-i Dunya—the roof of the world. Extending for nearly 1,800 miles, the Karakoram-Himalayan-Tibetan belt is a dynamic geological laboratory, formed by the closure of two ancient ocean basins and the collision of numerous blocks of Earth's crust. It holds an irresistible attraction for a structural geologist like me. I wanted to see the peak and walk the trails myself, and so I joined the Legacy Expedition led by Conrad Anker.

It's hard to describe the strong emotions I felt seeing Mount Everest for the first time. As the clouds briefly parted in the Khumbu Valley, Everest held her head high above the massive peaks in the foreground. Her summit pyramid was dark and rocky against the impossibly blue sky and, as from no other peak around, a massive plume of snow streamed easterly from her summit, a banner proclaiming to the world, *"I'm the tallest, I'm the goddess mother of all that lies below."* Then the clouds closed in again. My Sherpa guide (a kind and spiritual gentleman who has stood on the summit of Everest many times) smiled generously at me with empathy, or amusement, or perhaps both.

For me, the magic Mount Everest holds is the story written in her rocks. A geologist has to piece this story together from the fragmented manuscript that forms the landscape. The chapter that

**THIS 3-D COMPUTER** graphic (above) shows the Himalaya at an altitude of 24 miles (39 kilometers). To the left are the lush regions of lower Nepal and northern India; to the right, the high plains of Tibet.

**CONRAD ANKER (BELOW,** at left) and David R. Lageson, structural geologist, at Base Camp. Both men were members of the 2012 Everest expedition.

**WHILE THE SOUTHWEST** Face of Everest (above, at left) may be five and a half miles above sea level, it was in fact once the bottom of the ocean. A plate boundary collision between two continental landmasses pushed what would become Everest higher and higher, explaining the fossils of marine life that have been discovered on the peak.

**A DIORAMA AT** the Smithsonian's National Museum of Natural History (opposite) is a window into the world where Everest got its start. During the Ordovician period, which began about 490 million years ago, the sea was home to diverse marine life, including these cephalopods, corals, bivalves, crinoids, and trilobites.

most interests me is recorded in the layers of gray marine limestone that make up the summit pyramid. That's why I am drawn to the top.

## LIFE ON THE TETHYS SEAFLOOR

Standing at 21,300 feet (6,500 meters) and gazing up at Everest's southwest face in the brilliant afternoon sun, I watched the gray limestone of the summit pyramid turn a silvery gold. The Yellow Band, right below, also turned toward orange. The jet stream howled, sounding like the roar of a giant waterfall, but the only flow of water was beneath my boots—the slowly moving, frozen tongue of ice called the Khumbu Glacier.

Stumbling among the rocks near Camp II, out of breath and cold, I looked for lithic treasure. Sometimes one can find a small piece of gray limestone that has tumbled down some 7,700 feet from the summit to the glacier. These rocks tell a magnificent story of mountain building, from seafloor to summit, that spans nearly 500 million years of Earth's history—the story of how an ancient seafloor became the crown of the highest mountain on Earth.

Our journey begins some 480 to 470 million years ago, on the seafloor along the northern coast of the great supercontinent Gondwana, to which present-day India was attached. Major rivers, as big as today's Amazon, empty into the sea in broad deltas. Cumulus clouds build on a hazy afternoon horizon. No birds exist yet, but beneath the ocean's surface there are primitive, jawless fish that browse for food alongside clusters of tiny, slithering, wormlike creatures. The seafloor is rather flat, with a few shallow banks and canyons that funnel shifting sediment into a deep abyss far to the north. Elsewhere, fine-grained sediment drifts in among the fragmented remains of invertebrate animals, resulting in carbonate sediments that will form the limestone and dolomite of today.

## ORDOVICIAN INVERTEBRATES

This seafloor is part of the Tethys Ocean, an ancient body of water named for a Greek sea goddess. The patch of it in question we will call the Chomolungma shelf (or simply CS), after the centuries-old Tibetan name for Mount Everest. The seafloor at CS hosted a marine ecosystem teeming with life. If we followed a shaft of sunlight as it penetrated the upper marine layer, the sea shelf would have been revealed in dim, ever diminishing light. The water was somewhat cooler

"NO, OUR WORLD MAP IS NOT YET WITHOUT ITS SECRETS."

–JOSEPH F. ROCK, LEADER OF THE NATIONAL GEOGRAPHIC YUNNAN EXPEDITION, 1930

THE SPINY TRILOBITE fossil, once a scavenger on the seafloor, is a species that has been found on the peak of Everest. Trilobites were one of the most successful early complex life-forms, flourishing in the seas for about 300 million years. They had hard bodies and ranged in size from about an inch to over a foot long.

at these depths, but certainly not cold. Based on microscopic studies of rocks brought back from the summit of Mount Everest, we can infer that the seafloor at CS consisted not of clay mud but rather small pellets of carbonate mud excreted by a wide array of marine invertebrate animals. When these creatures eventually died, their exoskeletons contributed to the accumulation of carbonate sediment.

A large variety of trilobites scurried across this seafloor. Some spiny trilobites preferred to move about on the surface to scavenge food, whereas more robust varieties burrowed into the soft sediment to find their food. Ostracoda is a class of the Crustacea, sometimes known as "seed shrimp." Their tiny bivalve shells resembled a small clam, and most ostracods lived on the seafloor or within the first layer of sediment. Brachiopods, with their distinctive symmetrical shells, diversified markedly during this geologic period to form an important part of the bottom-dwelling community. Crinoids were also extremely abundant on the seafloor, forming vast communities of "sea lilies" attached to the seafloor by a flexible stem or stalk. Crinoids were filter feeders that used delicate, featherlike "arms" to capture their food from the warm oceanic currents that flowed along the bottom. Crinoids became so common that their skeletal parts formed incredible thicknesses of limestone. In addition, corals were found on the seafloor, since the period saw the ascendancy of massive coral reef communities in tropical oceans around the world. Stromatoporoids, a group of extinct sponges, also flourished in these reef communities. With all these species, it was a healthy, thriving, unpolluted marine ecosystem—a prolific biogenic factory of marine carbonate rocks that, someday in the far distant geologic future, would be uplifted to highest realms on Earth.

## THE GREAT BIODIVERSITY EVENT

These rocks were deposited during the Ordovician period, 488 to 444 million years ago, when life-forms diversified and became more complex—the most dramatic increase in marine biodiversity in Earth's history. Known as the Great Ordovician Biodiversity Event (GOBE for short), marine biodiversity tripled at the order, family, genus, and species level. The few seafloor-dwelling invertebrates that dominated were replaced by a diverse array of more complex marine animals during the GOBE.

Several geologic factors may have contributed. Four major continents existed during the Ordovician: Gondwana (to which India was attached), Laurentia, Baltica, and Siberia. The broad dispersal of these continental plates resulted in more surface area for tropical shelves around their margins. In addition, rising sea level across the globe contributed to flooding of inland seas, which resulted in more shallow marine habitats and increased carbonate deposits. Added to this, Earth

**MARINE REMAINS FOUND** high atop Everest demonstrate the geological history that this part of Earth has undergone, reaching back hundreds of millions of year ago to the Ordovician period, when the continents had entirely different shapes from today's.

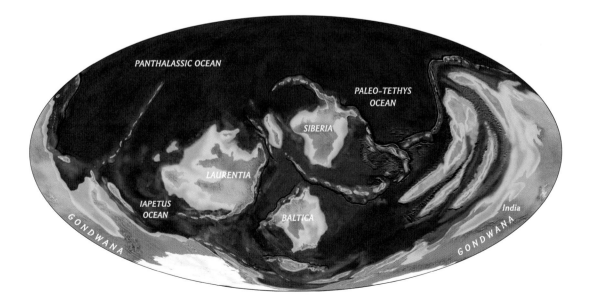

climates warmed during the early to mid-Ordovician, encouraging an increase in micro-phytoplankton, which may have provided a major nutrient source at the base of the food chain for marine invertebrate animals.

During the GOBE, a vast diversity of suspension-feeding and other organisms came to dominate the oceans and continued to do so throughout the remainder of the Paleozoic era. We see evidence of this atop Mount Everest, and it is somehow fitting that such a great evolutionary event is recorded by the highest outcrop on the highest mountain on Earth today.

## THE DISCOVERY OF EVEREST

One of the most ambitious scientific undertakings of all time occurred during the first half of the 19th century on the hot and humid plains of India. British explorers began the Great Trigonometrical Survey of India in 1808 with the ambitious task of determining the precise shape of the Earth, in addition to surveying the vast Indian subcontinent. Although the shape of the Earth had been determined in the mid-1700s by two French surveying expeditions, the precise degree of equatorial bulging and polar flattening remained a question. The British India Survey consumed the greater part of the century and employed some of the best mathematicians in the empire. The surveyor and mathematician in charge of this enterprise was William Lambton, a British Army officer who gained much of his early surveying experience in North America. Surveying equipment of the time was heavy and enormous, consisting of calibrated chains, measuring rods, and massive theodolites. Malaria, torrential monsoon rains, and tiger attacks tormented the British surveyors and their Indian staff, but they diligently carried on.

The Great Trigonometrical Survey finally reached the southernmost foothills of the Himalaya in 1847. Unable to gain access to the Kingdom of Nepal, surveyors were forced to work in the Tarai, a strip of land to the south, in order to survey the great peaks of the Himalaya more than 100 miles away. Andrew Waugh, surveyor general, measured distant "snowy peaks" from Sonakhoda, below the Darjiling hills near the eastern end of the Himalaya, in the autumn of 1847, after the monsoon season. One peak stood out among the others, now known as Kangchenjunga, the third highest mountain in the world at 28,169 feet (8,586 meters). Waugh did not announce his measurements for several more

THE BRITISH INDIA Survey, which began in 1808, was an extensive undergoing. This 19th-century lithograph (opposite top) shows Indian survey porters carrying some of the cumbersome equipment needed for the job: a tripod, a measuring chain, and a ten-foot-tall leveling staff.

A TIBETAN PRAYER wheel (opposite bottom) is used to spread spiritual blessings and well-being, but this one has been specially adapted with 100 beads, rather than the more typical 108. Pundits measuring distances used these adapted prayer wheels as a way to help count their steps. Two thousand steps measured approximately one mile.

SIR GEORGE EVEREST (left) was surveyor general of India for 13 years, and as a tribute to his work, the tallest mountain on the maps—first dubbed Peak XV—was given his name.

# "MOUNTAINS ARE AMONG THE LEAST KNOWN AND LEAST UNDERSTOOD AREAS OF THE WORLD."

–HUW LEWIS-JONES, BRITISH EDITOR AND HISTORIAN

years, in part because of another peak observed on the Nepal-Tibet border—a distant giant given the name Gamma (γ), from the Greek alphabet.

At the same time, in November 1847, a surveying assistant named John Armstrong also focused his surveying instrument on "γ" from a position farther west. He named the peak "b" and assigned to it a preliminary elevation of 28,799 feet (8,778 meters). Waugh distrusted the elevation calculations of "γ-b" and decided to wait for further measurements and recalculations before making an announcement. Over the next two years, Waugh sent two surveyors to the Tarai for additional sightings and measurements, but clouds and distance repeatedly thwarted their attempts to gain more data.

After several attempts, in 1849 James Nicolson finally obtained several vertical and horizontal angles from multiple stations closer to this mysterious peak than any surveyor had ever been; his preliminary calculations yielded an elevation of approximately 30,200 feet (9,205 meters). But Nicolson's initial calculations

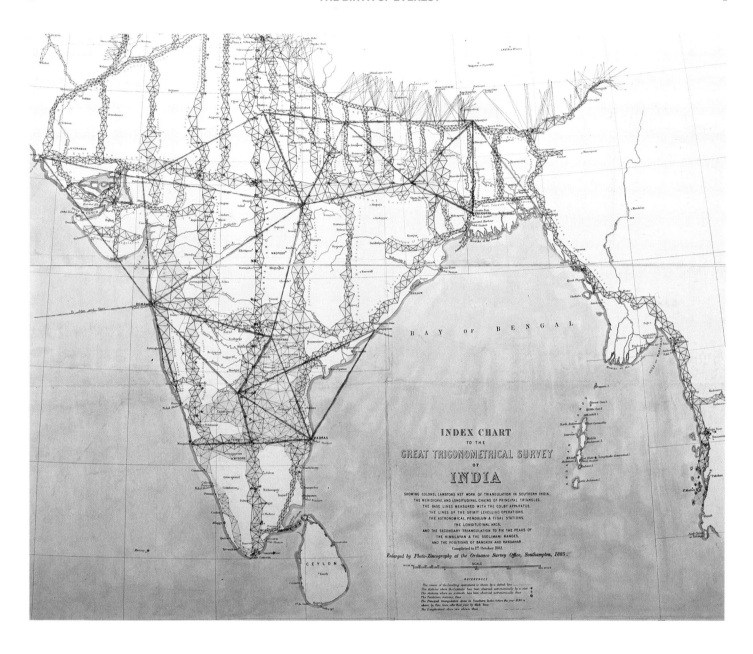

INDEX CHART
TO THE
GREAT TRIGONOMETRICAL SURVEY
OF
INDIA

did not take light refraction into account, which can introduce significant error into height calculations.

Several more years passed until Waugh's "chief computer" at the time, a brilliant Bengali mathematician named Radhanath Sikdar, most likely became the first to determine the first accurate elevation of the peak in question. In March 1856, Andrew Waugh finally made the formal announcement. In a 14-paragraph document, paragraph 5 reads: "We have for some years known that this mountain is higher than any hitherto measured in India and most probably it is the highest in the whole world." He named the peak Mount Everest, or Himalayan Peak XV, locating it at 27° 59′ 16.7″ N, 86° 58′ 5.9″ E and reporting its elevation as 29,002 feet (8,840 meters) above sea level.

British surveyors preferred to name Himalayan peaks with their indigenous names, if such names could be found, yet Waugh was unable to learn any local

**LAMBTON'S GREAT THEODOLITE** (opposite), used throughout the Great Trigonometrical Survey of India, could measure both horizontal and vertical axes but was less than portable. Weighing in at half a ton, it took 12 men to carry it. The 1870 index chart to the Survey (above) shows a network of triangulation, allowing for the first time an accurate framework for a map of India.

names, Nepal being closed to foreigners. He named the mountain after his predecessor, Sir George Everest (pronounced EEV-er-est by the family), who did not endorse the use of his name in this manner. Nevertheless, the name Mount Everest was subsequently approved in London by the Royal Geographic Society, and knowledge of Earth's highest peak rapidly spread around the world.

## ROOTS BENEATH MOUNTAINS

Scientific research on the geology and geophysics of great mountain ranges such as the Andes and Himalaya has a long history, with investigators not only trying to determine their origin but also attempting to answer basic questions such as: *What holds these mountains up? Is there a maximum height they can never exceed? How is rock mass distributed within them?*

The answers to these questions address the basic nature of how mountains are geologically assembled. Trekking from Lukla to Everest Base Camp over the course of many days, I was often struck by the tremendous significance of these great mountains to the historical development of the geosciences worldwide. Essential scientific ideas have originated over the past 150 years in the study of these lofty peaks.

One of the most impressive scientific outcomes of the Great Trigonometrical Survey of India was the discovery that mountain ranges have deep crustal roots that extend down into the Earth's uppermost mantle, analogous to the "roots" of an iceberg that extend well below sea level. Although 19th-century British surveyors in India knew that the mass of the Himalaya could influence the vertical angle of their survey plumb lines, a phenomenon known as astrogeodetic deflection, the geophysical details of why this happened were largely unknown.

George Airy, the Astronomer Royal, came up with one analysis of the plumb line deflection. He postulated that deep roots of low-density crustal rock—granite, for example—lie beneath great mountain ranges, and thus a high mountain range is "density-compensated" by its deep crustal roots. Today we know that the Himalaya does have a massive, low-density crustal root, produced by the continental collision of India with Asia. The continental crust beneath the Greater Himalaya shoots down 44 miles—double the average thickness of continental crust in the nearby lowlands.

## TWO EARLY GEOLOGISTS

At its best, science is an incremental process whereby new theories and ideas are built upon those of the past. Our understanding of how great mountain belts are created has evolved tremendously since the early development of plate tectonics in the 1960s. Many geoscientists deserve special mention; indeed, many historic climbing expeditions to the Himalaya have included one or more research geologists, including the 1963 American Mount Everest Expedition with glaciologist Maynard Miller. I have selected two geologists to highlight out of many distinguished Himalayan researchers. These two individuals are role models for younger geologists not only for their professional accomplishments but also for their strong character and spirit, which still echo among the great peaks and river gorges of the Himalaya.

### NOEL ODELL

Noel Odell (1890–1987) had a long and distinguished career as a British geologist and accomplished mountaineer. He served with the Royal Engineers, worked in the mining and petroleum industries, and taught geology at Harvard and Cambridge

NOEL ODELL (BELOW) was both a geologist and a mountaineer. He was a member of the British Everest expedition in 1924 (opposite), where he was the last person to see George Mallory and Sandy Irvine alive. Odell was one of the first to study the geology of the Himalaya, collecting samples on that 1924 trip of what he first thought were fossils.

universities, among others. Odell is perhaps most famous for being the last person to see George Mallory and Andrew "Sandy" Irvine alive as they attempted
to ascend the Northeast Ridge of Mount Everest on June 8, 1924. When Mallory
and Irvine failed to return to camp, Odell twice climbed without oxygen to around
27,000 feet (about 8,200 meters) in hopes of finding them. His ability to climb so
high was taken as proof for many years that the additional weight of supplemental oxygen equipment (called "English air" at the time) was detrimental to high-
altitude mountaineering. In 1936, he was the first (with Bill Tilman) to successfully
ascend Nanda Devi, which, at the time, was the highest mountain ever climbed.
Odell returned to Everest in 1938 as a member of the Tilman expedition.

During the 1924 expedition, Odell excitedly thought he had discovered fossils in a four-foot-thick bed of metamorphosed calcareous sandstone at about
25,500 feet (7,772 meters) on Mount Everest. Initially identifying them as casts
of lamellibranch fossils (bivalve mollusks), he later recanted this claim, showing the supposed fossils to be instead cone-in-cone, an inorganic structure fairly
common in clay-rich rocks. Odell also showed photomicrographs of indisputable

Noel Odell awoke early on the morning of June 8, 1924, and set off, up toward Camp VI, full of optimism for his friends George Mallory and Sandy Irvine, who were heading for the summit of Mount Everest. At about 26,000 feet, he looked up toward the highest reaches of the mountain. As he did so, the cloud that had been building since the late morning parted, affording him a view of the Northeast Ridge and the summit. In his diary he recorded: "At 12.50 saw M & I on ridge nearing base of final pyramid." He was the last to see them alive.

## STILL A MYSTERY

This last sighting has been minutely scrutinized and debated ever since, but no one has ever been able to prove that he did not see Mallory and Irvine so close to the summit. Odell's contribution to the 1924 Mount Everest expedition is often confined to this one observation, but actually his role on the trek and on the mountain was of far greater importance than history would suggest. For one thing, he turned out to be the man who best acclimatized to the altitude and was the only person who climbed above Camp IV to look for the climbers when they disappeared. For another, perhaps less obvious, he was a mentor for Sandy Irvine. He and Sandy had first been on expedition together in 1923 when they crossed the Norwegian island of Spitsbergen with a four-man sledge party. Sandy's performance gave Odell the confidence to recommend him to the Mount Everest Committee, who were seeking a "superman" for the 1924 expedition.

In fact, he and Sandy had met even earlier than that, in 1919, when Odell was on honeymoon in North Wales. He and his wife had been walking in the Carneddau, the northernmost range of hills in Wales, when they encountered an intrepid young motorcyclist on the top of Foel Vras, some 3,000 feet above the village of Llanfairfechan.

The motorcyclist approached the Odells and asked them for directions to the village. They pointed out the track and watched as he motored off across the rough ground and down the steep track. It was only when Odell retold that story in a tent on Spitsbergen that Sandy recognized himself and revealed that it was he who had been the motorcyclist. From that moment, a bond deeper than superficial friendship was established, and during his training for the expedition, Sandy wrote regularly to Odell, sharing his delight in everything he was doing.

Noel Odell at camp during the approach to Everest, 1924

## BACK TO THE TENT

After Odell had seen Mallory and Sandy going strong for the top, he went on to Camp VI and was amused by the tent, which was strewn with bits of oxygen apparatus. Sandy turned all his tents into workshops, and this was no different. Odell left some food and descended to Camp V.

Sandy Irvine and George Mallory never returned to their tent. They had died on the mountain, but of course Odell was unaware of this. He went back up to Camp VI the next day and again the following day in the hope that he would see some sign of his friends. We know he saw nothing, but few stop to imagine how sad he must have been when he realized he would have to break the news, first to climbing leader Edward Norton and then to the rest of the world, that the great George Mallory and his young climbing partner, Sandy Irvine, had perished so close to the summit.

Odell is far more than an historical cipher. He is central to the 1924 Everest mystery.

—JULIE SUMMERS  A British author and historian, Julie Summers has published ten books, including *Fearless on Everest: The Quest for Sandy Irvine.*

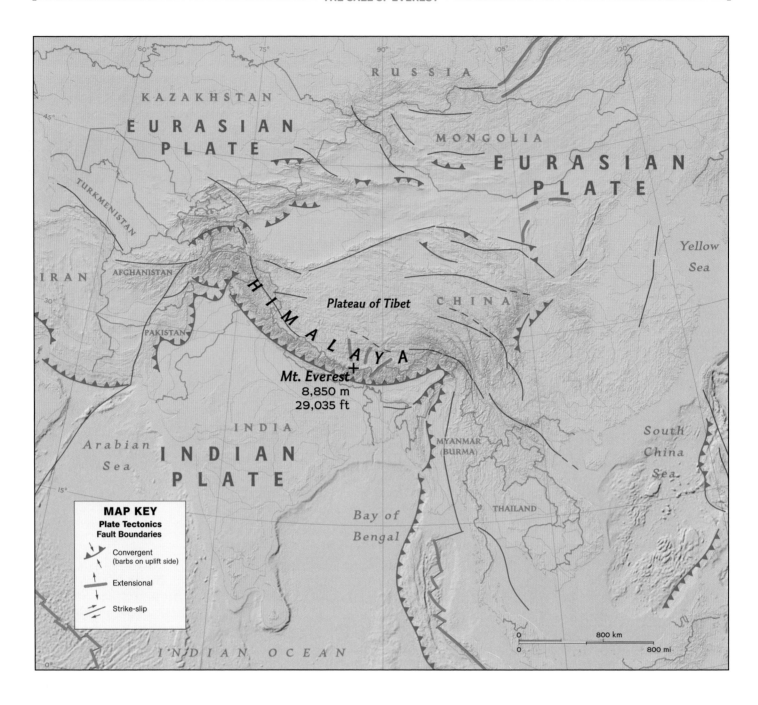

**WHEN THE SUPERCONTINENT** of Pangaea began breaking up, the Eurasian landmass and Indian subcontinent slowly drifted toward each other. When the two began colliding about 70 million years ago, the seabed was folded and raised, eventually leading to the uprising of the Himalaya.

crinoid fossils—leftovers of marine animals found in the summit limestone of Everest, confirming the observations of Augusto Gansser, who obtained summit samples from Swiss climbers in 1956 and the American team of 1963. Although Odell's initial discovery proved to be inorganic, his pioneering work on the geology of Mount Everest earned him great respect.

## AUGUSTO GANSSER

Augusto Gansser (1910–2012), a Swiss explorer and geologist, participated in the famous 1936 Swiss Himalayan expedition, crossing into the forbidden territory of Tibet dressed as a Buddhist pilgrim. He was one of the first Westerners to see the sacred Mount Kailash, headwaters of four major Himalayan rivers (the

Indus, Sutlej, Brahmaputra, and Ghaghara, which flows into the Ganges). He was impressed by the fact that the strata comprising Mount Kailash had been uplifted more than 20,000 feet (6,100 meters) and yet remained essentially horizontal, whereas surrounding rocks were highly tilted and deformed. Gansser also discovered the remains of ancient oceanic crust south of Kailash, which led to the naming of the plate tectonic boundary, or suture zone, between India and Asia. In this regard Gansser and his mentor, Arnold Heim, were decades ahead of their time in their interpretation of Himalayan geology.

Gansser's research took him back to the Himalaya for five field seasons in Bhutan, where he conducted more pioneering geologic work along the Tibetan border in uncharted mountains rising higher than 24,000 feet (7,300 meters). He published profusely, including the tome *Geology of the Bhutan Himalaya* (1983) and, most significantly, the definitive volume *Geology of the Himalaya* (1964), for which he produced a geologic map that spanned the entire length of the Himalayan-Tibetan

## THEN & NOW — RECORD KEEPING

| 1956 | 2012 |
| --- | --- |

A member of the 1956 Fritz Müller expedition makes observations by hand, while Travis Corthouts, in 2012, checks the weather from his laptop at Base Camp.

**SNOW CAN HARDLY** cling to the dark North and South Faces of Everest (at left). Although the summit of the mountain is composed of marine limestone, the rocks also carry the subtle signature of tectonic deformation and heat associated with the uplift of the Greater Himalaya.

region (from longitude 70° E to 95° E) at 1:2,000,000 scale—a masterpiece of geological synthesis and unimaginable hard work. Gansser was honored with the title of Baba Himalaya—Father of the Himalaya—by the University of Peshawar for his extensive pioneering work on the geology of the region.

## CONTINENTS COLLIDING

We now understand that the Himalayan mountain belt is the result of a massive collision between two landmasses, India and Asia, building on the theory of continental drift formulated in the early 20th century by the German scientist Alfred Wegener. Drawing from rock formations and the fossil record on either side of the modern Atlantic Ocean, Wegener hypothesized that a giant supercontinent existed around 250 million years ago, a supercontinent we now call Pangaea. As this supercontinent slowly broke apart and the continents "drifted" to their present positions, India slowly made its way north toward Asia as the Tethys Ocean basin closed ahead of it. As the last scraps of the oceanic crust were subducted into the mantle, India collided with Asia to uplift the Himalaya.

Wegener's work on continental drift was based entirely on the geology of continents, since the technology needed to explore ocean basins simply did not exist in his day. All of that changed in the years following World War II with the rapid increase in technology and our ability to explore space and deep oceans. What was found amounts to nothing less than one of the greatest scientific revolutions in the history of scientific inquiry—the birth of plate tectonics in the late 1960s. We now understand that the uppermost layer of the Earth is composed of a mosaic of semirigid plates that slide over a weak layer in the upper mantle and interact along plate boundaries. The north flank of the Himalayan mountains marks the boundary between the Indian-Australian plate and the Eurasian plate. GPS (global positioning system) satellite measurements confirm that even today, India is moving northeast at approximately 1.5 inches per year relative to Siberia.

The Himalaya took shape during a multistage collision involving several smaller terranes (or mini-plates). It is a story that takes us back to the floor of the Tethys Ocean, 480 to 470 million years ago, and the shelly carbonate sediment deposits on the Chomolungma shelf. Layers of sediment accumulated on the CS seafloor and became compacted by pressure, cemented with calcium carbonate, and recrystallized into hard limestone, entombing the invertebrate marine organisms as fossils in solid rock. As the loose collection of terranes inched northward, one ocean basin closed ahead through subduction—the Paleo-Tethys—while another opened behind through seafloor spreading—the Neo-Tethys. Southern Asia, including the present-day Tibetan Plateau, was assembled through three collisional events that added new crustal blocks from north to south, a process known as tectonic accretion—the dominant process by which continents have grown outward throughout Earth's history. When India eventually collided with Eurasia, a fourth major suture zone was created. This enormous collision weakened the lithosphere of southern Eurasia and, as a result, India was able to punch northward ever deeper. It has been estimated that India moved north into Eurasia more than 1,250 miles over the past 50 million years. This combination of partial subduction, thrust faulting, and folding of upper crustal rocks created the Himalayan mountains.

## BEDROCK ANATOMY OF MOUNT EVEREST

Topographically, Mount Everest is a massive pyramid with three great faces and three ridges leading to the summit. The Northeast Ridge extends three miles from the summit to a col, or gap, that separates the East Rongbuk and Kangshung Glaciers. A mile down from the summit on the Northeast Ridge, a shoulder marks the top of the North Ridge, which subdivides the North Face and separates Everest

> # "SPENDING TIME IN WILD PLACES WILL ALWAYS REMIND YOU THAT IN THE GREAT SCHEME OF NATURE, YOU ARE JUST THE TINIEST SMIDGE OF ALL THE OTHER ENERGIES AT PLAY."
>
> —STEPH DAVIS, AMERICAN ROCK CLIMBER

# THEN & NOW ALTIMETERS

## 1924

## 2012

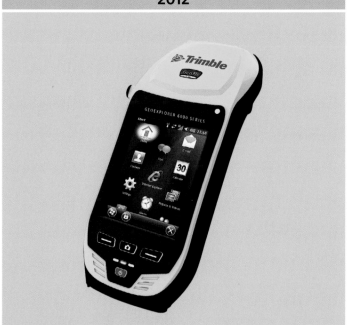

Pioneer climber George Mallory died on the mountain's icy slopes in 1924, and only in 1994, during Conrad Anker's expedition, were his body and broken altimeter discovered. Everest historians had hoped the antique device would indicate the altitude he reached, but its weathered works did not solve the mystery. Built to withstand harsh environments, today's Trimble GeoXH 6000 is a handheld GPS instrument designed for high-accuracy mapping and navigation. Built into it are the Windows mobile operating system and a camera.

from Changtse, or North Peak. The West Ridge, first climbed by Tom Hornbein and Willi Unsoeld on the 1963 American expedition, extends about three miles in a west-northwest direction from the summit to Lho La, a col between Khumbutse and Everest. The short Southeast Ridge, about one mile long, curves downward to the South Col, which separates Everest and Lhotse.

Between these ridges stand the three great faces of Everest: the North Face, the exceedingly steep Southwest Face, and the icy Kangshung Face. Each face has its own personality, created by the direction it faces, the hardness of its rock layers, the structural features that cut through, and the dip direction, or incline, of transecting rock units. The long Northeast Ridge is a succession of overlapping sedimentary rock layers dipping down the length of the ridge to the northeast. The First, Second, and Third Steps on the Northeast Ridge are formed by the down-dip ends of rock layers, resting on layers below like overlapping roofing shingles. On the shorter but steeper Southeast Ridge, the dip of rock layers is "into the mountain," thus creating blunt, steep steps like the South Summit and Hillary Step.

The bedrock geology of Mount Everest is decidedly more complex than its topography. Everest is not a layer cake of rock units; it is not an eroded remnant of horizontal strata like those exposed in the walls of the Grand Canyon. On the contrary, Everest is dissected by major faults that juxtapose different types of rock. Any color photograph of Mount Everest shows four distinctive intervals of rock differing in color and layer characteristics. From top to bottom:

- Gray limestone makes up the summit pyramid;
- The Yellow Band, the most distinctive band of rock on Everest, encircles the peak like a gold wedding ring;
- Dark gray, thinly bedded, fine-grained, weakly metamorphosed, sedimentary rock of the Everest Series is broadly exposed below the Yellow Band; and
- White granite and high-grade metamorphic rocks (gneiss and schist) form the lower portions of the mountain.

## SUMMIT ROCKS

Forming the highest outcrop on Earth are sedimentary rocks consisting of gray, finely laminated, silty limestone and dolomite, first identified as ancient marine sedimentary rocks by Noel Odell and described by Augusto Gansser as

**EVEREST'S YELLOW BAND,** at a slight slant around the peak, is actually interlayered dolomitic marble, fine-grained phyllite, and semi-schist rock. Climbers know they've reached the band when their crampons hit the hard rock, which requires several hundred feet of rope to traverse.

fine-grained, thin-bedded, gray "platy limestone" or "calc-schist," distinguished by the presence of crinoidal fossil fragments. Gansser used these fragments to support his contention that the summit limestone was late Paleozoic in age, around 300 million years old, although he recognized that fossil preservation was not good enough to determine an accurate age. More recent geologists have noted other invertebrate fossils from summit limestone samples, and a distinctive white-weathering, 200-foot-thick thrombolite bed—common microbial deposits found in ancient marine carbonate rocks—has been described at the base of the Third Step on the Northeast Ridge. These fossils, in addition to stratigraphic studies in southern China, establish an Ordovician age for the summit limestone on Mount Everest: roughly 470 million years old.

## YELLOW BAND

The distinctive Yellow Band is arguably the most recognizable feature of Mount Everest, forming a high ring of yellow-tan marble around the peak right below the summit. I like to think of it as a gold wedding ring. The Yellow Band lies immediately below the Chomolungma detachment fault and crops out below the South

Summit on the Southeast Ridge and below the First Step on the Northeast Ridge. About 650 feet thick, it is a succession of interbedded dolomitic marble and phyllite, a fine-grained metamorphic rock with a silky sheen, suggesting that these rocks were subjected to higher temperature and pressure than the overlying formation. The Yellow Band is believed to be Middle Cambrian in age, about 30 to 40 million years older than the Ordovician rocks of the summit pyramid.

## EVEREST SERIES

Across most of the three faces of Everest, below the Yellow Band, lies the Everest Series, primarily low-grade metamorphic rocks derived from mudstone or shale, including metamorphosed sandstone and minor thin beds of marble. Common minerals found in these rocks include albite, chlorite, epidote, biotite, and quartz. These rocks likely represent deep-water sediments, mostly mudstone and clayey sandstone, deposited on the continental shelf somewhere north of Gondwana. A major river system may have been the source for the sand, silt, and clay in the original rocks, much like the modern Mississippi River, which dumps its load of muddy, silty sediment into the waters of the Gulf of Mexico.

## GREATER HIMALAYAN SEQUENCE

The base of Mount Everest is composed of igneous and metamorphic rocks that form the metamorphic core of the Himalaya, a giant slab of rock called the Greater Himalayan Sequence, easily observed along the trekking route from Lukla to Everest Base Camp. Dark metamorphic gneiss, a banded rock similar to granite, stripes the walls of deep gorges cut by the Dudh Kosi River. On the steep switchback trail leading up to Namche Bazar you notice swirled migmatites, a metamorphic rock that was once partially molten. The stone steps cut into the near-vertical hillsides between Namche Bazar, Phortse, and Tengboche provide fresh exposures of the metamorphic minerals and fabrics. As you approach Ama Dablam and beyond to the lower Khumbu Glacier, the proportion of white granite in boulders along the trail increases significantly. At Base Camp the landscape becomes monochromatic, a black-and-white vertical world of white granite and ice. This is the exhumed core, the innards, of the mountain belt—the upper crust of India that was detached, metamorphosed, and thrown back on itself to the south to form the Greater Himalaya.

## SOUTH TIBETAN DETACHMENT SYSTEM

Two great faults, the Chomolungma and Lhotse detachments, are exposed across

"IT'S A REGION OF ASTOUNDING BEAUTY, FULL OF HISTORY AND MYTH. I DON'T KNOW ANY OTHER MOUNTAIN LIKE IT."

—KENTON COOL, BRITISH MOUNTAINEER

the Southwest Face of Mount Everest. Together they are called the South Tibetan Detachment System. In geologic terms, a fault is a fracture in the Earth's crust that results in a demonstrable offset between the types of rock to either side of it. The narrow, brittle Chomolungma detachment lies at the top of the Yellow Band and carries the summit pyramid in its hanging wall. The wider fault zone of the Lhotse detachment lies between the granites and gneisses of the Greater Himalayan Sequence and the overlying Everest Series. North of Everest, these two faults merge and extend east-west for at least 400 miles.

Geologists have determined that the Greater Himalayan slab was uplifted and tectonically unroofed through simultaneous motion on a major thrust fault at the base of the slab (exposed over 30 miles south of Everest) and the South Tibetan Detachment System at the top of the slab (in the Everest region). Called channel flow tectonics, the process is something like squeezing toothpaste from a tube, and the Greater Himalaya were literally extruded southward between these two great fault systems. What a story to ponder as one trudges in the rarefied air above the

**THE MOON RISES** over Nuptse on the Nepali side of the Himalaya. The mountain is known for its thick masses of leucogranite, a light-colored granite. "Spider Wall," on the south face of Nuptse (at left), is actually a network of leucogranite dikes.

Western Cwm to higher regions on Everest! One is not merely climbing a mountain but climbing through the uppermost crust and across a major fault system that has literally detached the roof of the world from its deeper roots below.

The relatively new model of channel flow tectonics has revolutionized our thinking about the structural geology and tectonic evolution of mountain systems. We now understand that great mountains not only wear away through the slow but relentless surface processes of glacier and stream erosion; they can also become tectonically eroded through large-scale detachment faults that literally decapitate the highest terrain.

Three factors contribute to the modern-day uplift of the Himalaya:

1. Continued collisions between the Indian and Eurasian plates at a rate of about 1.5 inches per year;
2. Tectonic unroofing and erosion of the Himalaya; and
3. Relentless erosion of the southern aspect of the Himalaya caused by the annual summer monsoon from the Indian Ocean.

WHEN THE EURASIAN and Indian plates collided, it marked the birth of many great things: the highest mountains in the world, the highest plateaus, and the largest concentration of glaciers outside of the poles. And while geologists continue to study how this region came to be, it continually changes.

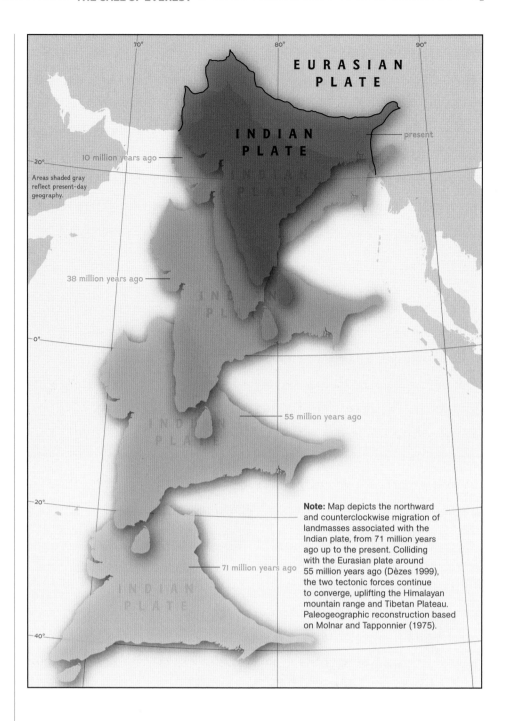

EURASIAN PLATE

INDIAN PLATE

present

10 million years ago

Areas shaded gray reflect present-day geography.

38 million years ago

55 million years ago

**Note:** Map depicts the northward and counterclockwise migration of landmasses associated with the Indian plate, from 71 million years ago up to the present. Colliding with the Eurasian plate around 55 million years ago (Dèzes 1999), the two tectonic forces continue to converge, uplifting the Himalayan mountain range and Tibetan Plateau. Paleogeographic reconstruction based on Molnar and Tapponnier (1975).

71 million years ago

INDIAN PLATE

The fact that climate, through long-term precipitation and erosion, can influence tectonic process operating miles down in the crust is a revolutionary concept. In this way, mountains are master teachers of geologic knowledge, and the Karakoram-Himalayan-Tibetan belt is truly one of the greatest geologic laboratories on Earth.

I can think of no more fitting tribute to the poetry of Earth's tectonic engine than the fact that this region, now called the top of the world, was once a tranquil seafloor, a subtropical habitat for the invertebrates that thrived in the warm, clear waters of the Tethys Ocean. Today modern rivers carry sediment, dissolved ions, and nutrients from the roof of the Himalaya back to the Indian Ocean, just as rivers did on Gondwana more than 400 million years ago. Perhaps some calcium ions, dissolved from the limestone on Everest, will make their way down the great river systems to the Indian Ocean and be used in the construction of a new calcium-carbonate shell of another seafloor creature. Perhaps this little creature will find its way to the top of a new mountain range formed by the eventual closure of the Indian Ocean as another continent crashes into India in a couple of hundred million years or so. This is how it works and how it has always worked with the Earth: a grand circle of recycling and evolution that includes both the organic and inorganic world, interwoven through deep geologic time.

## A MOUNTAIN IS BORN

Of the Karakoram-Himalayan-Tibetan belt, we can cite a number of geologic superlatives: It contains the highest mountains on Earth (14 are higher than 8,000 meters, just over 26,000 feet), the greatest relief of any continent, the highest uplift rates, the largest concentration of glaciers outside the polar regions, the world's highest and biggest plateau, and the source of many of the world's greatest river systems. Little wonder that this region has attracted geologists since the early days of exploration and surveying. We can see the evidence of past collisions on continents around the world, but the enormous elevation of the Himalaya, coupled with deep erosion, provides unparalleled opportunities to study the inner workings of an active, evolving collisional mountain belt, a veritable tectonic work in progress.

Mount Everest is a magnificent testimony to the power of nature. It boggles the mind to think that, through the complete rearrangement of Earth's continental and oceanic plates, ancient ocean-floor sediments could be lifted up into the summit of the highest mountain on Earth over the course of 400 million years. As I stood at the foot of Mount Everest in the Western Cwm, the mountain seemed everlasting and solid, built on a base of granite and capped by a crown of limestone. Chomolungma, the goddess mother mountain, will surely endure all that nature throws at her, just as her summit parts the jet stream with a flowing banner of snow.

But Chomolungma is only a temporary mark on the landscape, like everything in geology. Chomolungma is an unfinished geologic symphony. She will continue to morph and change through time, perhaps growing higher before eventually succumbing to erosion, gravity, and glaciers. It's impossible to predict when, but another Himalayan peak will eventually assume the title of roof of the world, and then another after that, and another after that. Such change is inevitable on a dynamic, living planet. Change is why we exist in the first place, and why we have the privilege to ponder it all.

# "THE HIMALAYA ARE STILL GROWING AT ABOUT THE RATE YOUR FINGER-NAILS GROW."

-BRADFORD WASHBURN, CARTOGRAPHER AND PHOTOGRAPHER

# CHAPTER 3

# THE PEOPLE

Broughton Coburn

# OF EVEREST

A Buddhist monk plays a drum during the Mani Rimdu festival at Thame monastery in Sagarmatha National Park, Nepal.

# IN THE FALL OF 1984

I TRUDGED NORTHWARD ACROSS A BROAD, FLAT SNOWFIELD. ITS HIGHEST POINT AROSE IN THE FORM OF A STOUT WOODEN PILLAR, MARKING THE BORDER BETWEEN NEPAL AND TIBET. THIS WAS THE LEGENDARY NANGPA LA, THE 19,050-FOOT PASS LOCATED IN THE SHADOW OF CHO OYU, A PEAK THAT RISES TO THE WEST OF MOUNT EVEREST.

In the fall of 1984 I trudged northward across a broad, flat snowfield. Its highest point arose in the form of a stout wooden pillar, marking the border between Nepal and Tibet. This was the legendary Nangpa La, the 19,050-foot (5,800-meter) pass located in the shadow of Cho Oyu, a peak that rises to the west of Mount Everest. I tied a *kata* prayer scarf to a tangle of others on the post and recited Buddhist mantras common to Tibetans of the Tibetan Plateau and to the Sherpas of Nepal to the south. Retracing my steps, I caught a fragrance that blew in gently from the south: pine and juniper forests warmed by the sun, melded with dust and incense and wood smoke from the lowlands of Nepal and the Gangetic Plain of India.

Descending on the Nepal side of the border, I followed a maze of granite rubble that had been churned up and deposited on the glacier's surface. Yaks can follow the scent left by previous caravans, but lone travelers must decode their way through the monochromatic scree. I found no sign of yak dung—every scrap had been thrown onto the loads of the yak trains for use as cooking fuel. And the cairns meant to mark the trail were just as the Tibetan traders had described them: imaginary piles of stones that moved and disappeared like phantoms.

**SHOPPERS SPEND TIME** looking for goods at Saturday's open-air market in Namche Bazar (above). Perched on a steep mountainside, the Sherpa town is important for the local economy. Ethnic Sherpas (such as these giggling children below) share cultural roots with Tibetans and live in Nepal and Darjiling, India.

**YOUNG BUDDHIST MONKS** enacting a ritual of protection (above)

**THREE YOUNG MONKS** stand beside the Rinpoche at Tengboche Monastery in 1956 (opposite). The Rinpoche was in his early 20s when Fritz Müller took this photograph. Now 78, he remains a spiritual leader of the Khumbu region.

The route over the Nangpa La has long been a conduit of history, trade, and culture for Tibetans and their Sherpa kin—the people of Everest. I was entering Khumbu, the Sherpas' homeland—a region circumscribed by the headwater valleys that drain the Nangpa La and the south side of Mount Everest. Khumbu's waters eventually merge into the Dudh Kosi River and flow toward India. One monk described the outline of Khumbu and its veins of rivers as "shaped like a flower bud about to bloom."

Nearly 500 years ago, Sherpa legend says, hunting dogs belonging to a man named Kira Gombu Dorje chased a wild sheep across the Nangpa La. The dogs, in turn, were pursued by the hunter—the first Tibetan to settle in Khumbu's fertile, temperate valleys. The south side's greener pastures soon drew a pioneer community of *shar-pa*, "people of the east," who became citizens of Nepal.

In winter, when farming and herding is relatively dormant in Tibet, people from the north side of Everest still climb over the Nangpa La. They are either traveling on Buddhist pilgrimage or as refugees from China, seeking opportunities in India or the West. Their bid to reach the outside world—by scaling the spine of the Himalaya—underscores the Tibetans' irrepressible desire to experience the freedom that the Sherpas to the south enjoy.

## THE SPIRIT OF KHUMBU

The Sherpas of Khumbu barely number more than 3,500, with a parallel population of 17,000 more in Solu, a region three days' walk to the south. For at least a century these two enclaves, distinguished by subtle differences in dialect and dress, have been closely linked by marriage. From the mid-19th to mid-20th centuries, the arcane laws of Nepal's Rana oligarchy restricted trade with Tibet to men who lived in border districts. This made Khumbu husbands especially desirable to brides from Solu, by allowing their families to marry into a valuable trade connection.

But the Sherpas' connection with Tibet was maintained as much for religion as it was for trade. Beginning in the early 1900s, many of the Khumbu Sherpas sent their boys north, back across the Nangpa La, for monastic training at the Buddhist monastery of Rongbuk, near Everest's Base Camp on the Tibet side of the mountain. From this monastery the Sherpas drew much of their formal religious traditions.

# "SPIRITUALITY AND RELIGIOUS DEVOTION HAVE BECOME INTERTWINED WITH OUR DAILY LIVES."

–LODI GYALTSEN GYARI, SPECIAL ENVOY TO THE 14TH DALAI LAMA IN THE U.S.

CHINA
NEPAL

**CHINA**

Mt. Everest
8,850 m
29,035 ft

Kathmandu

Namche
Bazar

N E P A L

BHUTAN

28°

27°

85°     86°     87°     88°     89°

I N D I A

BANGLADESH

0     100 km
0     100 mi

Kathmandu
1,349 m
4,426 ft

**FEBRUARY 20, 1963**
"A little war against a big mountain"
begins with the assembly of the American
Mount Everest Expedition in Kathmandu,
Nepal's capital. Thence the expedition
jounces by jeep and truck to Banepa.

Bhadgaon

An army of a thousand men,
including climbers, Sherpas,
porters, and a Nepalese liaison
officer, sets out from Banepa
on the 185-mile hike to Everest,
February 20.

Panch-khal

Banepa
1,555 m
5,102 ft

Dolalghat

Risingo

Chitare

Chyaubas

Kirantichhap

Sun Kosi

☐ Camp site

0     10 km
0     10 mi

**A TYPICAL TREK** by foot from Kathmandu to Everest Base Camp can take six weeks, as did the 1963 American expedition, mapped above. Today vehicles and airplanes carry trekkers to Lukla, almost 85 miles beyond Kathmandu, which reduces the journey time. Trekkers must still plan for time spent acclimating along the way.

Tenzing Norgay, who climbed Everest with Ed Hillary in 1953, grew up in the Tibetan village of Kharta, directly to the northeast of Everest. After studying briefly at Rongbuk, he crossed the Nangpa La, in the 1920s, and settled with his parents in the Khumbu village of Thame. At age 18, he left Khumbu to search for work in Darjiling—a hiring and staging point for foreign climbing expeditions to Everest.

Tenzing's nephew Nawang Gombu—who climbed Everest with Jim Whittaker in 1963—also studied at Rongbuk for two years, until he could no longer endure the strict monastery regimen. One night he bundled up his clothes, sneaked out through the hole of a latrine, and walked over the Nangpa La to Khumbu.

The Cultural Revolution began in 1966—seven years after the uprising against the Chinese in Lhasa. During that chaotic period, the Rongbuk monastery was destroyed. Fortunately, most of Rongbuk's arcane and colorful Buddhist traditions had already regerminated, with a distinctive flair, and were thriving in the Sherpa monasteries of Solu and Khumbu.

MAY 1
James W. Whittaker, with Nawang Gombu, becomes the first American to attain Everest's summit. Four other Americans reach the same goal, May 22.

*Cho Oyu*
8,188 m
26,864 ft

CHINA
NEPAL

Mount Everest
8,850 m
29,035 ft

Gorak Shep
Base Camp

Lobujya

*Nuptse*
7,861 m
25,790 ft

*Lhotse*
8,501 m
27,890 ft

*Makalu*
8,485 m
27,838 ft

Helicopter picks up frostbitten William F. Unsoeld and Barry C. Bishop at Namche Bazar and flies them to a hospital in Kathmandu, May 27.

Pheriche

*Ama Dablam*
6,812 m
22,349 ft

Pangboche

*Imja Khola*

Thyangboche
MARCH 9-14

*Bhote Kosi*

Namche Bazar
3,350 m
11,000 ft

MARCH 7-8

Phakding

Lukla

Yersa

FEBRUARY 27

*Likhu Khola*

*Bhent Khola*

Puiyan

*Dudh Kosi*

Those
1,905 m
6,250 ft

Chyangma

MARCH 2

Taksindhu

Kharikhola

*Inukhu Khola*

Sete

Junbesi
3,048 m
10,000 ft

Jubing

*Khimti Khola*

# SACRED SITES

Below the Nangpa La lies no ordinary mountain valley. I was now descending into a *beyul:* one of several "hidden valleys" of refuge designated by Padmasambhava, the ninth-century "lotus-born" Indian saint, revered by the Sherpas as Guru Rinpoche, who subdued the often wrathful territorial deities of the mountains and converted them into defenders of the Buddhist faith.

In order to protect Buddhist beliefs and teachings during times of adversity, Guru Rinpoche's female consort is believed to have hidden "treasure texts" in the valleys' forests and hillsides. These sacred documents are time capsules, essentially. When revealed in future eras, devout people will be able to recover their faith and relearn the wisdom of the ancients.

These beyul are blessed with spiritual energy, and the Sherpas say that one should behave with reverence when passing through this sacred landscape. Here the karmic effects of one's actions are magnified, and even impure thoughts should be avoided.

Sites of spiritual power are scattered throughout Khumbu: ancient meditation caves, "self-emanated" handprints and footprints impressed in rock, and curious snakelike intrusions that for Sherpas represent the *lu* serpent spirits. Small shrines and bamboo wands, adorned with prayer flags, are set up at springs, wells, and stream confluences—places graced by the nourishing flow of water. Villagers say that the rain from the sky and the water in their springs can dry up if these sites are defiled—for instance, if a lowlander were to slaughter a goat nearby and dress the meat in the water source. This offends the lu and can cause them to flee.

Khumbu's harsh environment only seems to invigorate the Sherpas' faith. Just as the Himalaya were formed by intersections and accretions of geologic material, the religion of the Sherpas is accretionary—a blend of shamanism, pre-Buddhism, and Bon and Tibetan Buddhism—while pantheons of clan gods and territorial deities lurk and cavort on the hillsides.

## THE YETIS

In 1974, a woman named Lhakpa Drolma spotted a yeti while herding yaks along a stream in a valley above Khumjung. In her account, it advanced quickly toward her and attacked, knocking her out. She awoke, half submerged in the stream, and found three of her yaks dead—with their heads missing. Lhakpa Drolma still becomes edgy and fearful when people around her speak of the yeti.

"Yeti" derives from *ya-te,* abominable animal of the glaciers. It is surpassed in fearlessness, though, by the *mhi-te,* a long-haired, dwarf-size humanoid that eats people. The Sherpas say that even a glance from the mhi-te—especially if viewed from below—can cause illness or possibly death, and even saying the names of these dreaded creatures should be avoided. Another variety of dwarf hominid is known as *min-jung-tale.* It is said to have a monkeylike face and short, stubby feet. A necklace of human ears (or bones) graces its chest, and it uses a femur as its walking stick.

Some Sherpas say that the yeti are associated with a class of deities known as the Dharma Protectors and may be emanations of a wrathful cemetery goddess called Du-to Lhamo, and perhaps other deities. Yetis are not always dangerous, however, and are sometimes said to be playful tricksters, prone to petty thievery.

The Tengboche Lama suggested that one expedition looking for yetis in the 1950s leave behind a camera so that he could take a picture. "They don't seem to

**BUDDHISTS PLAY THE** gyaling, a double-reed instrument, as a young boy looks on (opposite top). Gyalings, which sound similar to an oboe, are wind instruments in the Tibetan musical orchestra. The white scarf from the mouthpiece to the bell is a Buddhist offering scarf known as a kata.

**THE INCARNATE LAMA** of Tengboche in about 1953 (opposite bottom). While only about 10 percent of Nepalis are Buddhist, the religion has deep roots in the region. The Lord Buddha, Siddhartha Gautama, was born in 623 B.C. in Lumbini in the foothills of the Himalaya. Sherpas belong to the Nyingmapa or old sect of Mahayana Buddhism.

# THEN & NOW TOURIST DESTINATION INTACT

### 1924

### 2012

Equipment and technique have radically changed over the near century of Everest expeditions, but Kathmandu's Durbar Square, hub for enthusiastic tourists, retains its historic flavor and spectacular architecture that includes a royal palace dating to the 11th century.

The snow falls white now, but soon that brightness will fade." These words echoed in my mind after my audience with the Rinpoche, head lama of the Tengboche monastery. The old man's eyes looked weary with time and his efforts of selflessness. He sits in his simple quarters, studying the scrolls containing the written history of his people, the Sherpa. After almost 70 years as the community and religious leader of Khumbu, he says he is becoming weak of body and mind.

"All of the lamas believe in true Sherpa culture and with them, the ways of our ancestors are preserved. If the lamas ever lose the belief and trust of the Sherpa people, though, our culture will be lost." —Rinpoche

The Tengboche monastery perches on a remote and wind-blown hill, high in the Himalaya, shadowed by the vast immensity of Ama Dablam. To most Westerners, this place and the monks residing within embody the ancient and now revered Buddhist culture of the high Himalaya. Hand-hewn stone walls, elaborately painted, tell colorful stories of the daily rituals of lamas, praying and playing ancient instruments that resonate through the mountain valley.

"Tourists come here for our culture as it once was, so that they may feel that they have ventured into the unknown, but the truth is that the face of our younger generations holds only the surface of what our culture once was." —Rinpoche

One cannot fault the young people of the Khumbu region for moving away from the old ways of life. Taking the path of modern education and technological advances over a simple life of physical labor has become a trend no matter where you look. But what has been lost in this process, in the modernization of one of the most remote places in the world? It is a

Four Sherpa women in traditional Nepali garb sit in attendance for a Sherpa baby-naming ceremony in Namche Bazar. All family members and close and distant friends are invited to such celebrations.

question I ask in my wanderings through the Khumbu region, this once mysterious and isolated realm of mountain gods.

## DRAWN TO KHUMBU

In 1990, when I was two, my father, Alex Lowe, traveled to Khumbu for the third time and became the 40th American to summit Mount Everest. Ascending Everest was still a far-fetched and seldom accomplished feat at that time. I learned of the titanic mysticism of these settings via postcards and the stories he told me as I lay in the crook of his arm. From early on I felt the draw of the Himalaya—so much so that in a childish temper tantrum, I declared to my mother, "I am running away to Kathmandu to find Dad."

My father died in 1999, consumed by an avalanche in the Tibetan Himalaya. The region gained even more grave and mystic proportions for me. In 2001 I was finally able to see Nepal through my own eyes when I accompanied my mom and my new dad, Conrad Anker, to the Sherpa climbing school that they had launched in Phortse, Khumbu, Nepal. And now again, in the spring of 2012, I return to work on my research and photography project, looking at this remarkable culture and geography through the eyes of its long-standing inhabitants, people I now consider friends and family.

To many people, Nepal and the Himalaya are still remote and mysterious places, to be visited only by the most bold of travelers. In the last 50 years, though, trekking through the Khumbu and Annapurna regions of the Himalaya has become a huge tourist industry, one of the largest sources of income for the economy of Nepal. The first power plant was established in the region in 1995, and with it came running water, telephones, the Internet. From the top of Mount Everest, the highest peak on Earth and one of the harshest environments in the world, you are now able to make a cell phone call.

In recent years, a dental clinic and several medical clinics have opened, run by Sherpa practitioners who left for an education and returned to work among their people. Modern clothing and comfortable homes provide a better standard of living, while money and easy access via air and road to the lowlands has diversified the once elementary diet of potatoes, rice, and the few vegetables that grow in such a harsh climate. Life has become far easier and more comfortable for the Sherpa people of Khumbu, and they are grateful for this.

## FOR BETTER, FOR WORSE

If you ask people living in Namche and nearby to name any negative effects of Western culture, seldom do you get an answer. Here, as everywhere, people seem to prefer a lavish, easier life with Western-style amenities. "The snow falls white now, but soon that brightness will fade," says the lama.

Several days later I was able to speak with Lama Geshi at the monastery in Pangboche, just a day's hike up the valley from Tengboche. We spoke of what the Rinpoche had told me, and Lama Geshi closed his eyes, bowed his head in understanding, and began to speak.

"Hundreds of years ago, the religious kings of Tibet enforced the practice of self-giving and selflessness, and this is where the happy disposition of the Sherpa people comes from. People lived simple lives here in the mountains without the knowledge of the swirling storm around us. Westerners came, bringing with them money and the idea of power. Now the people no longer focus on self-giving and the idea of a simple life. . . .

"I am 80 years old now, and I have experienced the impermanence of human life. All the most powerful and rich people in the world will die, just like everyone else, so live each day to better others. Through this, you will better yourself and find true fulfillment. All the people in the world should remember that they will be dead one day, tomorrow, next week, or in 50 years. Live for the day and live for a settled mind. Don't waste days planning ahead when the days spent planning could be wasted days of doing." —Lama Geshi

The clarity of what Lama Geshi said rang lucid in my mind. With the access to endless monetary gain, people might better their own existence, but to what end? Western culture and influence has most certainly had a positive effect on the quality of life for individuals, but the collateral loss may not be worth the cost.

"Stopping a flood is impossible, and these days it rains heavy from the east and west." —Lama Geshi

—MAX LOWE **A National Geographic Young Explorer Grantee and photographer, Max Lowe is documenting social change in Nepal's Khumbu region.**

STUDENTS LINE UP at Mount Kailash, a boarding school in Kathmandu, Nepal. These students' parents pay for their education, a luxury in a country where rural students can walk for miles over rugged mountain trails to reach a schoolhouse. Nepal's literacy rate is only 60 percent for citizens older than the age of 15.

come out when foreigners are around," he explained. Sherpa trek leaders make sure that the legend of the yeti remains alive among clients—by dramatically rattling their tents on occasion, once the trekkers settle into their sleeping bags.

## THE MOUNTAIN GODDESS

One of these deities resides on Mount Everest: Miyolangsangma, the Goddess of Inexhaustible Giving. The mountain is her palace and playground, and Sherpas view themselves and the climbers as guests at Base Camp, arriving without invitation. Only when the Sherpas have made offerings to Miyolangsangma and other deities at the *lhap-so*, a makeshift shrine that is normally built soon after arrival at Base Camp, do they feel comfortable entering the treacherous Khumbu Icefall. Miyolangsangma is regarded as one of the five "Long Life Sisters" who reside on nearby peaks of the region, protecting the area from harm while providing spiritual nourishment to the residents who dwell on her flanks.

In paintings, Miyolangsangma is depicted riding on a tigress that levitates her through a celestial realm. She exudes an air of stern benevolence, as if requiring that her beauty and generosity be reciprocated with respect and offerings. A cornucopia of fruit in her right hand represents good fortune, wealth, and abundance, and her left hand is positioned in a gesture of giving. It is her power, one monk said, that has delivered to the Sherpas great bounty—in the form of climbing expeditions and foreign travelers, to begin with.

Like many of the Sherpas' deities, she simultaneously occupies a position of endearment, fear, and devotion. In 1953, Tenzing Norgay credited the goddess with granting him safe passage on the slopes of Everest, and for escorting him and Edmund Hillary to the summit.

## MODERN LIFE

The 1950s and early '60s marked a socioeconomic watershed for the Khumbu Sherpas. Traditional trade with Tibet, over the Nangpa La, was curtailed by the new Chinese regime, and Khumbu's villages were languishing in a less than vibrant barter economy. Sherpa women, mostly, cultivated potatoes and weaved blankets, then carried them down valley to trade for rice and corn.

"My sister and brothers lost hair on the tops of their heads from the constant

# "HOW COULD ANYONE NOT FEEL THE URGE TO GIVE SOMETHING BACK?"

—JON KRAKAUER, MOUNTAINEER AND AUTHOR OF *INTO THIN AIR*

# VOICES THE PUSH AND PULL OF PROGRESS

Many Sherpa youth regard the older generation as backward because they are skeptical of new technology, underestimate the value of education, and reject any new understanding of the position of women. Young women are critical of the fact that they are brought up more conservatively than the boys and tied to the home, while the boys are allowed more free time and are given the chance to go abroad. In the Sherpa culture, it is the Sherpanis [female Sherpas] who are the preserving force and who guarantee continuity and stability while the men are away. The girls' upbringing stems from the perception of the women's role as caretakers and perpetrators of their culture . . .

Language and religion are undisputed core values for the Sherpas. Myths and religious rituals structure the pattern of their year—a feature that not even the most modern young people would like to change. Buddhism and its values are unquestioningly upheld.

This cannot be said for the Sherpa language, however. Many Sherpas speak better Nepali than Sherpa. . . . The educated young people are aware of the danger that Sherpa culture could disappear altogether if core values such as language are not upheld, and they demand protective measures to save such traditional features. At the same time they are an integral part of a process of change.

—KURT LUGER **Kurt Luger is chairman of Eco Himal, an international NGO dedicated to sustainable development in the Himalaya.**

A YOUNG SHERPA plays pool in a pool hall decorated with a Bob Marley poster.

chafing of their tumplines," says Ang Rita, an educated Khumjung villager. "No one wore shoes, even in winter, though the better-off families sometimes tanned and shaped crude boots from buffalo hides."

In 1959, Khumbu had virtually no schools, drinking water systems, year-round bridges, or improved medical care. Mainly from experience as traders—which requires some math skills and the ability to draft business agreements—elder Sherpas had begun to recognize the value of education.

When Sir Edmund Hillary visited the region that year, six years after his summit triumph, he was approached by a quorum of villagers who presented him with a scroll-like petition. "Our children have eyes, but they are blind," an elder declared in one of several speeches delivered by the gathered Sherpas. They requested that he build a school. Hillary established an organization called the Himalayan Trust, and in 1961 the Khumjung school was constructed. Its first headmaster, from Darjiling, went door to door, persuading parents to send their children to the new school. The 46 students in the first class spanned a range of ages; most had no foundation of schooling at all.

"After two years of school," Ang Rita recalls, "my parents were surprised that I wanted to continue studying. 'Haven't you learned everything there is to learn yet?' they said. I was determined to keep going, so in the early mornings and late evenings I did all my chores—mostly climbing trees to chop firewood and carry it back to the village. Then I studied."

Over the following decade, Hillary's organization supported the construction of 22 more schools. As for Ang Rita, he went on to earn the highest score in Nepal in the School Leaving Certificate examination; he is now the director of the Himalayan Trust.

## HEALTH CHALLENGES

Khumbu needed a hospital, too. Cretinism, caused by iodine deficiency, was almost epidemic, and some villagers had goiters the size of small footballs hanging from their necks. Alcoholism was common as well. Ulcers, complications from childbirth, TB—all were present, though not readily apparent to casual visitors. Then, in 1963, smallpox found its way into Khumbu from India. If not for the quick response of the Himalayan Trust—and the doctors of the 1963 American Mount Everest Expedition—the Sherpa community would have been harder hit. Porters carrying loads to Everest were found to be

**SHERPA PORTERS DESCEND** from Everest's South Col after leaving supplies for another team ascending. A Scottish physiologist, A. M. Kellas, first recognized the Sherpas' high-altitude abilities in the early 19th century when he traveled in the Himalaya and employed Sherpas in his expeditions.

**SHERPA PORTERS WORKING** during the 2012 climbing season pose at Base Camp with various loads. By assisting in early climbs, Sherpas gained a reputation for their deep understanding of the mountain and their physical abilities to operate in an oxygen-starved environment.

vectors of the disease, as they were more mobile and contagious than most villagers. In the spring of that year, vaccine was flown in and the trust was able to immunize almost everyone in the region, including the people in Solu, to the south.

In 1966, a hospital was opened in the village of Kunde. An iodization campaign was promptly launched, and virtually all new cases of goiter and cretinism were eliminated. Still, illnesses continued to be blamed on the influence of ghosts and malevolent spirits, and even today many Sherpas consult shamans before going to the hospital—a place where wounds can be treated, but where people have also been known to die.

For four decades the hospital was staffed by husband-and-wife medical teams from New Zealand and Canada. It is now operated by Western-trained Sherpa doctors and staff. The Himalayan Trust, American Himalayan Foundation, and other organizations provide ongoing support.

## FURTHER EDUCATION

Hillary wanted to see Sherpas trained in all fields. In the 1970s, the government of New Zealand sent six Sherpas to New Zealand for graduate training in parks and natural resource management. All of them returned, and in succession three of the six were appointed as chief wardens of Sagarmatha National Park—which was gazetted in 1977 and encompasses all of Khumbu. Their legacy of enlightened park management carries through to the present day.

The generosity of foreign travelers, Hillary had realized, was leading to a culture of dependency. "If people want to assist the Sherpas and show their affection for them," he counseled, "I've always recommended that they contribute to the community as a whole, rather than to individuals." He found that the most sustainable projects were those in which the Sherpas invested their own time and labor, effectively making the enterprise their own.

Fortunately, schools and education arrived in Khumbu just as the Sherpas began to need tools for harnessing the socioeconomic transformation that was about to sweep over them.

The modern era for Khumbu began, effectively, outside of Nepal in the 1920s and '30s. Tenzing Norgay and his nephew Nawang Gombu were members of the first and second waves of Sherpas to seek their fortunes in Darjiling, India. Until 1950, when Nepal opened to outsiders for the first time, all Everest expeditions

# THEN & NOW  HIMALAYAN SCHOOLS

| 1956 | 2012 |
|------|------|

In 1953, Edmund Hillary and Tenzing Norgay became the first climbers known to reach Everest's summit. Soon after, Hillary founded the Himalayan Trust to build schools and hospitals. In 1966, schoolchildren learned their letters using crude wooden pencils and red sand. Today, the foundation's Western influences—from paper and pens to dress—can be seen.

passed through India and Sikkim, then into Tibet, to approach the mountain from the north side.

The Sherpas who immigrated to Darjiling effectively left their barely subsistent, pastoralist Khumbu relatives behind. They were on a fast track to a mercantile and monetized realm that the foreign climbers—the sahibs—commanded. For decades in the middle of the last century—1920 to 1980—the upwardly mobile Darjiling Sherpas moved comfortably between the old and new worlds. Their Khumbu relatives treated them politely but chafed at what appeared to be favoritism toward the Darjiling Sherpas. (Since then the Khumbu Sherpas have more than caught up, socioeconomically, though a mild rift continues to this day.)

Changes in Sherpa culture accelerated in the early 1960s. During the 1963 American Mount Everest Expedition in particular, Sherpas and Americans alike experienced a mutual loss of innocence. The Americans were nonplussed by the generosity and good humor of the Sherpas, and by how their lifestyle flowed in harmony with their environment. The Sherpas were equally amazed by the exotic Westerners with their profligate spending habits and command of technology. East meets West in the shadow of Everest.

In a milestone that went unnoticed at the time, Jimmy Roberts and a Sherpa named Nima Tenzing inherited the lightweight expedition gear from the 1963 expedition. The next year, Roberts registered a company called Mountain Travel—inventing a whole new industry for Nepal and the Sherpas called "trekking." The concept wasn't well understood at first, but within a decade more than 10,000 people had come to "trek" in the Himalaya. Many more have followed, and Sherpas have adapted to a new livelihood based on guiding, feeding, and lodging them. The Sherpas have gone on to open hundreds of such trekking companies, and many more hundreds of lodges. Now at least 25,000 foreigners trek through Khumbu each year, and the approach route to Everest offers reliable satellite TV, cell phone coverage, and eight-page restaurant menus.

## ROMANTICIZING THE SHERPAS

Academic studies of the Sherpas abound. When '63 American Everester Jim Lester returned to Nepal in 1998, his guide told him, "When you were here 35 years ago, every climber had his own Sherpa. Now every Sherpa has his own anthropologist." Some have studied the relationship between Sherpas and Westerners, which is charged with fascination, envy, and romantic notions. Visitors to Khumbu often describe their trek or climb as life-changing, and their local hosts

# "THE SUMMIT IS NOT THE END OF THE STORY."

—ANG RITA SHERPA

AT 3 A.M., smoke rises in the air as Danuru Sherpa burns juniper before climbing the Khumbu Icefall to fix ropes for the summit climb. Sherpas believe that the smoke from burning juniper helps to cleanse the climbers as they pass through it.

MINGMA SHERPA AND Apa Sherpa rest while fixing the ropes and ladders needed for the climbing route through the dangerous Khumbu Icefall. Known as the Icefall Doctors, Mingma and Apa fix and maintain the route throughout the climbing season. It's dangerous work—43 Sherpas died on the job in the first 70 years of climbing Everest.

form a key element of that experience. Sherpas have gamely followed the natural progression from noble savage to renaissance men and women, traversing a staggering cultural arc in less than two generations.

In the broader ethnic and social hierarchy of Nepal, Sherpas were regarded (until recently) as second-class citizens. Contact with Westerners and income from trekking presented them with refreshing opportunities to move upward and outward, and many have effectively leapfrogged over the socioeconomic structure that defines the rest of Nepal. Remarkably, this tiny ethnic minority has become famous amid a score of much larger ethnic groups.

## RISKY BUSINESS

Historian Audrey Salkeld stresses that Sherpas have paid a disproportionately high price in lives lost on Everest. In 1922, seven Sherpa porters were buried under an avalanche on Everest's North Col, and in 1974 six were swept away by an avalanche in the Western Cwm. In the first 70 years of Everest climbing, 53 Nepali and Indian Sherpas were killed—more than a third of the total climbing deaths in that period. Because of their contribution to route fixing have seen a lot of changes in Nepal since when I was young. We had a traditional way of living that was simple. The forest was intact. There was no disturbance from outsiders. But when Hillary climbed Everest in 1953, everything changed. The flood of tourists has put pressure on the local population, which is only about 3,500 Sherpa. And the pressure is not only environmental, but also cultural. That's why I chose to study park management.

Linking culture and environment together is a strong message. To preserve culture and environment it is important to create systems that benefit both. I think the best way to do that is to make sure that local people feel and have ownership for their resources.

Ang Rita Sherpa and his wife, 1992

## MANY MORE VISITORS

Between 1966 and 1970, the number of tourists quadrupled to 46,000, and by 1976 it had passed 100,000. Since 1997, annual tourist arrivals have averaged well above 400,000. Over 50 percent of these visitors go to protected areas.

Tourism in Khumbu, the Everest region, has had a wide variety of impacts. With the closure of the northern border, the Sherpas' centuries-old trade with Tibet has been replaced by tourism and international climbing expeditions as a source of cash income. Tourism has created new jobs and opportunities; it has boosted the living standard of local communities with better health care, education, and infrastructure.

On the negative side, tourism has shared much blame for accelerating environmental problems in the mountains. Garbage produced by trekking and mountaineering in the Everest region poses a significant environmental problem, and has become highly publicized. This is more than an aesthetic issue, as environmental pollution is also involved—not least with the indiscriminate dumping of old batteries, and nonbiodegradable wastes.

## PARKS PROTECT THE CULTURE

We have nine national parks, and eight of them are managed by the army. In Makalu-Barun National Park we did the opposite—we only hired staff from the community.

Since they are from there, they have so much knowledge about the area. And we found that we could put a stop to poaching if we recruited poachers' sons and daughters into the Scout program. Suddenly, everyone has a stake in the park that they want to protect.

With The Mountain Institute (TMI) our goal with local communities is to provide the example of what can be done. We ask communities to list their needs in a community consultation: school support, infrastructure, et cetera. Then we help them see that to gain something you have to work and explore your options. All we need to do is guide them. Then they will feel empowered to act in the future.

I developed a concept in 2003 called the Sacred Sites Trail. Instead of having all the tourists simply travel in a straight line from Namche to Tengboche and back again as it said in their 15-year-old *Lonely Planet* guide, I wanted all the communities to have a chance to benefit from tourism. We need to think through every aspect of the project in order to make it successful.

—ANG RITA SHERPA  The senior program manager of The Mountain Institute's Asia (Nepal) Program, Ang Rita Sherpa works on ecotourism and sustainable development.

SHERPA WOMAN WITH prayer
beads at Tengboche monastery

and ferrying supplies, especially in the Khumbu Icefall, Sherpas are exposed to riskier parts of the mountain than their employers. Now expeditions are required to carry a $4,000 life insurance policy for every Sherpa who enters the icefall.

If a Sherpa dies on Everest, the body is brought to a low ridge near 15,000 feet called Chukpo Lare ("rich man's yak corral") for cremation. Most expedition Sherpas have relatives who were cremated or memorialized here, and they always stop to recite a prayer for their benefit.

"Climbing is exciting but dangerous," a young Sherpa named Lhakpa says. "It's best left to young, single men. We don't gain much spiritual merit when climbing, unless we act selflessly in some way, or save someone's life." Like many, Lhakpa plans to build a lodge and invest in the "bigness"—business—end of trekking.

## PAST WAYS

In 1963, Sherpas led lives vastly different from today. "Most Sherpas bathed only once a year—at the start of the monsoon, during a two-week period known as *Maal-chu*," recalls Sange Dorjee, who now lives in Wyoming. "Before the summer festivals we washed virtually everything—clothes, personal items, and our bodies. It had a wonderful purifying effect." Maal-chu is followed by the festival of Phang-ngi, in June—when the *naks* (female yaks) and cow-yak crossbreeds produce the richest, most delicious milk, which is churned into *phangmar,* an exceptionally rich butter.

Kancha, of Namche—a veteran of the 1953 and '63 Everest expeditions—recalls making countless trading trips with yak trains to Tibet. Since then the flow of people and goods has reversed direction. "Sherpas used to endure hardship to make a small amount of money from the Tibetans. Now poor Tibetan traders climb over the Nangpa La to earn a small amount of money from us." But trade between Khumbu and the Tibetan Plateau may have entered a long slide. Sherpas can get all the (iodized) salt they need from India. And the Tibetans are connected by a paved road to the rest of Tibet, and China.

There are barely enough yaks in Khumbu to form a trading caravan, anyway. The word for "yak" used to be synonymous with the term for "wealth," and the families who owned the most yaks were the richest and most respected people in Khumbu. In an about-face, they are now the poorest. The youth needed to herd

**SHERPAS PAY CAREFUL** attention to their headgear in the summer festival of Dumje at Tengboche monastery. Buddhist monks wear the *tse-sha*, a saffron-colored hat, while laymen don cowboy-styled hats.

yaks have realized that foreign visitors are easier to care for than livestock, and tourists don't need attention year-round.

"Sherpas fear that the day will come when we no longer raise yaks," says Ang Rita. "When I was young, 16 families from my village—Khumjung—raised yaks. Now only three do." It's no longer profitable to keep yaks for seasonal load carrying, or for milk. Most milk now comes from outside Khumbu; powdered varieties from Holland and Denmark are preferred.

Many young Sherpas have forgotten—or aren't keen to speak—their tribal languages. In Sherpa homes in the cities, Nepali, English, and Hindi are becoming more frequently spoken than the native Sherpa tongue. The culture, and attachment to it, remain strong, but so are opportunities and distractions. Partly through cinema and satellite television, Sherpa youth have seen extraordinary, beguiling new worlds. The existence of these bountiful lands has been confirmed by newly rich relatives' tales of odysseys to the West.

Education and wealth mean that many Sherpas can hire lowlanders to till their fields and harvest their barley and potatoes. They no longer need to carry

**PACK ANIMALS SLOWLY** cross a frozen stream near Base Camp. Even if they slip on this ice, yaks are able to keep their balance. Both commerce and adventure have traditionally relied on these beasts of burden.

loads, guide trips, or even run their trekking lodges. Increasingly, they lease their wayside inns to low-landers of ethnic groups with names such as Rai, Tamang, and Gurung—many of whom are perfectly happy to quietly adopt the higher-status "Sherpa" name as their own.

## PRESENT CHALLENGES

In a development that is slowly tugging at the fabric of their culture, Sherpas have come under pressure to sell their family tracts to outsiders. Nepal forbids discrimination in the sale of land, so the Sherpas are obliged to honor reasonable offers. Preserving Sherpa cultural heritage in this context is difficult, however, because the lowlanders have no historic link to the deity-animated landscape, to the valley's sacred sites, or the forests and pastures that have been stewarded for generations.

Modern life has introduced modern challenges. Ngawang Karsang, a former doctor at Kunde, links some of the current diseases and afflictions to changing socioeconomic conditions.

"In the 1960s and '70s, alcoholism was common," he says. "Expedition Sherpas would drink before the expedition to get their courage up, then drink after the expedition to celebrate." Recently, however, he has noticed higher incidences of obesity, diabetes, heart attacks, hypertension, and strokes. Packaged, processed foods have found their way into Khumbu—indeed, into many high Himalayan valleys.

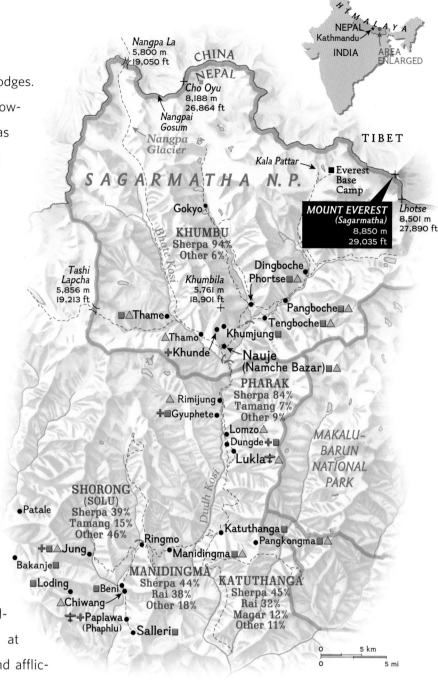

Himalayan Trust project
✛ Airstrip
⌒ Bridge
✚ Hospital, health clinic
■ School
△ Buddhist monastery, temple
-- Trekking route
KHUMBU Sherpa region with ethnic breakdown

# PRESERVING SHERPA WAYS

Some Sherpas, concerned by these losses of tradition, have made laudable efforts to preserve them. Former park warden Lhakpa Norbu has written a book documenting Sherpa ceremonies and material culture, and he is co-authoring a

**SHERPA PEOPLE LIVE** in Nepal and India. Those in the Khumbu region are most intimately connected with Mount Everest.

VIVID COLORS AND high peaks
fill a painting by Ang Temba
Sherpa (previous pages). A deaf
artist, Temba Sherpa received a
scholarship from the Himalayan
Trust Scholarship.

VIVID COLORS AND high peaks
fill a painting by Ang Temba
Sherpa (previous pages). A deaf
artist, Temba Sherpa received a
scholarship from the Himalayan
Trust Scholarship.

PANURU SHERPA OF Phortse in
Nepal's Khumbu region served
as sirdar, or head Sherpa guide,
for the 2012 North Face/National
Geographic expedition, which
commemorated the 1963 American ascent. Sherpas estimated
that some 300 people started
their summit pushes on May 18
or 19, the first of two windows of
good weather.

Sherpa language dictionary with the Tengboche Lama. Likewise, a young man named Lhakpa Sonam has turned a traditional house, furnished with original brass and wooden utensils, into a museum—even though many similar houses remain in active use. Other Sherpas have established funds and foundations that channel support for preserving monuments.

Meanwhile, in a village in Japan, an exact replica of the Takshindu monastery has been constructed by five Sherpa carpenters; the Tengboche Lama and a group of his monks traveled there to consecrate the structure and perform the traditional Mani Rimdu dance-drama ritual.

Khumbu has enjoyed a fragile cultural and environmental balance for decades. But the late Trulshig Rinpoche, abbot of Rongphu monastery in the 1940s and '50s, long ago predicted that much attention would be focused on Everest, and that people would "suffer hardship as a result of negative deeds generated in her vicinity." The Tengboche Lama, too, has suggested that the noise and the fumes of helicopters can disturb the deities of the hidden valley, which can result in calamities. Helicopter and plane crashes occur with greater frequency—though this could be attributed to the greatly increased air traffic.

Yet the land and people are remarkably resilient. Some of those who deeply respect the Buddhist traditions also regard the conservatism of the high lamas as antiquated in the modern age. At the same time, the monasteries benefit from the wealth that tourism and new enterprises have generated in their communities. Remarkably, the enrollment of monks and nuns has not declined over the past few decades.

## SHERPAS IN AMERICA

The first Sherpas to visit the United States came on a State Department–sponsored goodwill tour that was tacked on to the 1963 American Mount Everest Expedition. The five wide-eyed high-altitude Sherpas began their trip at the White House, where President Kennedy awarded them (and the American team members) National Geographic's Hubbard Medal.

A generation later, descendants of several of the 1963 Sherpas have settled in America, where they are thriving. Typically, Sherpas land first in Queens, New York. They drive taxis, wash dishes, or pump gas—entry-level jobs on a socioeconomic ladder formed of ambition and hard work. A number now work as commercial airline pilots, doctors, and businesspeople.

Everest gave the Sherpas a towering head start. But few have returned to Nepal, which still suffers an unstable political and business climate. Over time, this one-way emigration trend may reverse as new opportunities arise in the natural- and tourism-resource-rich Himalaya.

Already a handful of Sherpas have questioned whether relocating to the U.S. is the right choice. "I think some of us are becoming too much like Americans," says Daya Yangji Sherpa, a resident of Jackson Heights. "Now divorce has happened among Sherpas here—and it can be bitter and expensive, with lawyers getting involved endlessly. It's crazy."

To weather the special challenges—many of them financial—that go with life in America, a group of expatriate Sherpas bought and renovated a church in Queens. It houses a social welfare organization called the Sherpa Kyidug and provides a gathering place (online and in Queens) and cultural tent pole for American Sherpas. (Kyidug is said to be a combination of *kyi*, happiness, and *dug*, suffering—referring to weddings and funerals, the milestone events that mark the principal times the American Sherpas are drawn together.) Still, some of the

> "USUALLY WE THINK THAT BRAVE PEOPLE HAVE NO FEAR. THE TRUTH IS THAT THEY ARE INTIMATE WITH FEAR."
>
> –PEMA CHODRON, AMERICAN AUTHOR AND ORDAINED BUDDHIST NUN

SHERPAS AT BASE Camp are all smiles as they watch a movie after a day spent working to support the climbers. Sherpas summit the mountain in higher numbers than other climbers, and they are also more likely to have multiple summits.

elders and middle-aged talk of returning to Khumbu to retire—rather than die in a foreign land.

## TRADING PLACES

Ching Drolma is a Namche Bazar innkeeper with an animated smile. Like many middle-aged Sherpa women, she was a part-time farmer and yak herder and now works day and night tending to trekkers' needs. As she pounds away on a slab of male water buffalo meat with a wooden mallet—tenderizing it into shape to become a "yak" steak—she pauses. A gleeful smile creeps onto her face.

"I have an idea," she announces buoyantly. "All of you Americans seem to like our mountains and our houses and culture. Why don't you all come here and live in our houses—and we'll all move to America and live in your houses?"

This has been suggested before, but the line is always good for a laugh. This time, though, she is serious. She is sharing a recognition that we are all part of a global community, with few insurmountable boundaries. She is also underscoring how we all tend to idealize the world in which the other lives.

Standing at the *lhap-so*, our makeshift Buddhist shrine at Everest Base Camp, I thanked Miyolangsangma, the goddess who lives here, for welcoming me into her home. Then I headed upward through the icefall to meet the members of our expedition at Camp II. Bounding across ladders over deep crevasses, I noticed ice screws loosening in the late May sun. On Everest, sunshine is a blessing and a curse. The sun's warmth is welcome, but it causes the ice to melt. Large seracs crumble, overhangs avalanche, and pinnacles topple.

We had spent almost eight weeks waiting for the summit window, the longest delay I had ever experienced. Climbers had lost weight and muscle strength, and they spoke increasingly of changing weather patterns and global warming. Perhaps Miyolangsangma, weary of playing the perennial hostess, needed a rest.

For Sherpas, Miyolangsangma is the traditional supplier of food and wealth. But by allowing us to climb Everest, she has provided us with more than this. She has given our people education, medicine, and opportunities to work, and she has opened our eyes to a world beyond our valley. Without her blessings and without mountaineering expeditions, we would have remained isolated in our Khumbu villages with little means of improving our future.

## A FICKLE GODDESS

At Camp II, I heard the summit weather forecast for the coming days: 50 mph winds, snow, and a windchill factor of 20 degrees below zero. It would require another two days of climbing for our expedition members to be in place for a summit attempt and, given the weather predictions, it did not look favorable. Miyolangsangma, however, often changes her mind.

Two days later we climbed to the South Col, Camp IV, the last stop before the summit. The team members rested fitfully in tents blown nearly flat by a freezing wind, while I peered

Tommy Heinrich and Apa Sherpa on Everest, 1995

out at clouds that broke like ocean waves across the peaks below us. It would be a cruel and unforgiving night—the worst I had seen high on the mountain. A few hours after sunset, the wind calmed—slightly—and my team, along with other expeditions' climbers, departed Camp IV for the summit.

I prayed to Miyolangsangma for guidance and could hear the oxygen mask-muffled chants of Dawa, a former monk, also requesting permission to continue. In 1990, when I first climbed Everest with veteran climbers Rob Hall, Gary Ball, and Peter Hillary, Miyolangsangma welcomed me into her lap. Her invitation, however, is temporary.

Sitting on the Balcony, the responsibility of making the decision to continue up, or to turn around, weighed heavily on me. Yangji had urged me to come on this expedition with our friend John, to "keep him safe." Now I was caught by indecision—by a feeling that Miyolangsangma might be withdrawing her invitation.

As if sensing my dilemma, John gently reached a hand over to me and said, "Apa, are you okay? Are you warm enough?" The wind carried his words away, but I could see in his eyes that he was feeling strong, ready to push on. We would continue to the South Summit and, if conditions didn't improve, we would turn around.

Two hours above the Balcony, an amber dawn revealed the triangular shadow of Everest stretching to the western horizon—Miyolangsangma's welcome mat. It was now clear that we were invited guests and not interlopers.

John, Dawa, and I ascended the exposed slope to the South Summit, where we again switched oxygen bottles. We scrambled up the Hillary Step, traversed the final snowy humps, and, at 6:35 a.m. on May 31, John, Dawa, and I stood on the summit. I thanked Miyolangsangma for her hospitality once again. Then the three of us began the journey to our true destination: safe return to our families.

—**APA SHERPA** **Assisting Western climbing expeditions since 1985, Apa Sherpa made his first of 21 summits of Mount Everest in 1990.**

# CHAPTER 4

# THE NATURE

Alton C. Byers

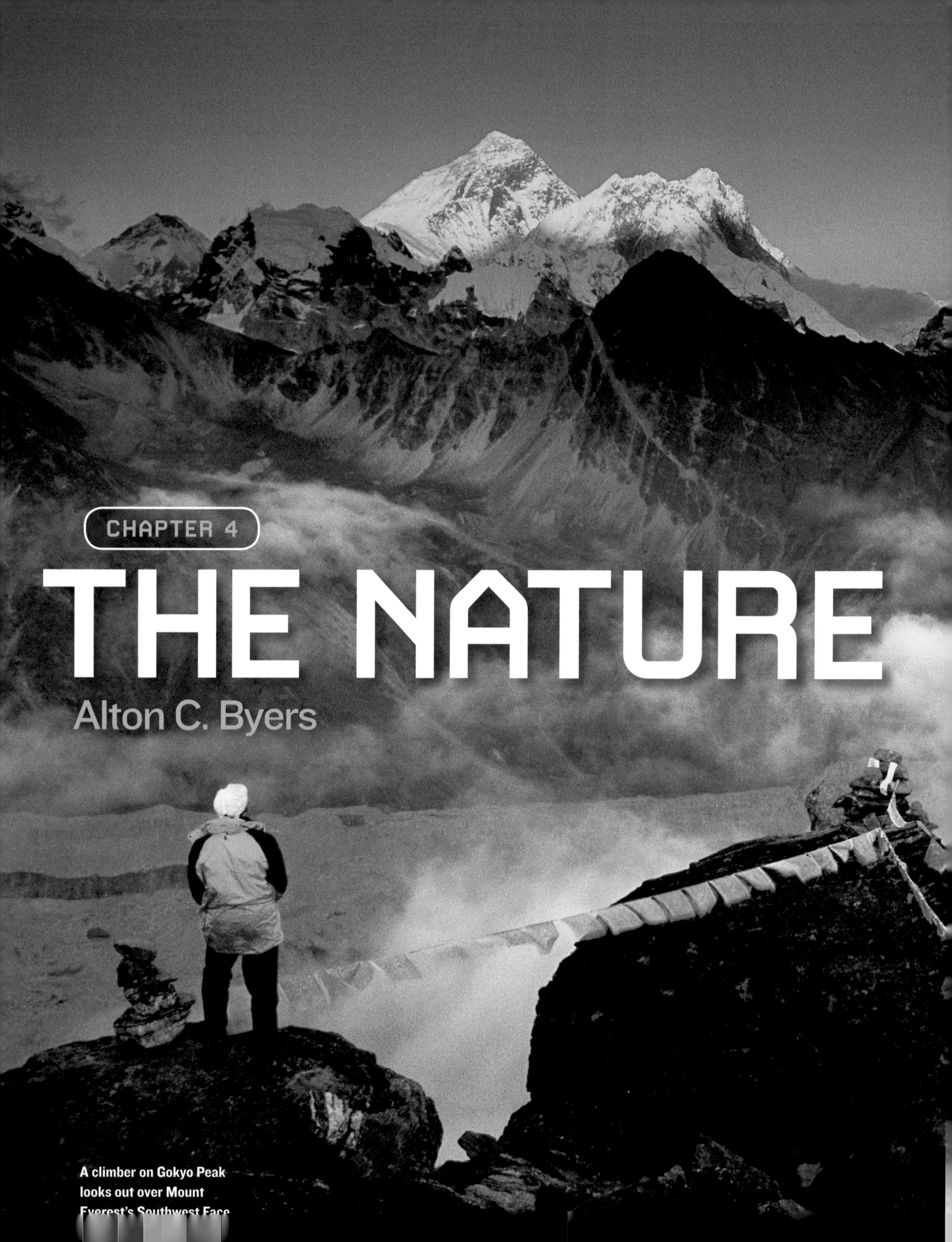

A climber on Gokyo Peak
looks out over Mount
Everest's Southwest Face

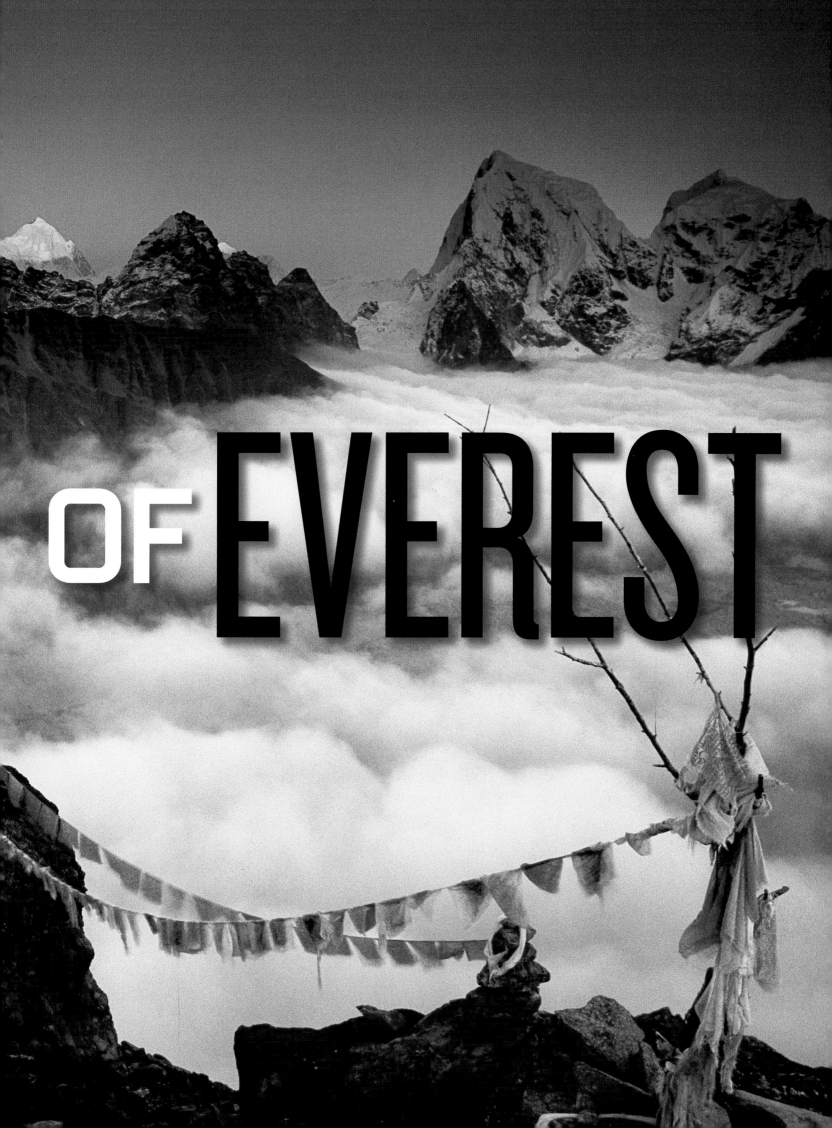

# ON A BEAUTIFUL, CLEAR MORNING IN MAY 2012, I HEADED THROUGH A FOREST HIGH ABOVE THE IMJA *KHOLA* (RIVER) IN THE SAGARMATHA NATIONAL PARK. RARELY USED BY TREKKING GROUPS OR VISITORS, IT WAS A TRAIL I DISCOVERED YEARS AGO WHILE LIVING IN THIS REGION OF NEPAL, KNOWN AS KHUMBU.

On a beautiful, clear morning in May 2012, I headed through a forest high above the Imja *khola* (river) in the Sagarmatha National Park. Rarely used by trekking groups or visitors, it was a trail I discovered years ago while living in this region of Nepal, known as Khumbu. I was spending a year conducting the fieldwork for my Ph.D. in geography. Once a week for ten months I would trek the 30-mile round-trip from our home in Khumjung village (at an altitude of 13,025 feet/3,970 meters) to the seasonal alpine village of Dingboche (at 14,500 feet/4,420 meters). As I

**AN ENDANGERED SNOW** leopard triggers a remote camera trap. With only a few thousand left in the wild, these nocturnal cats are a rare sight in the Everest region, but their numbers are increasing thanks to intelligent park management.

**GNARLED TRUNKS IN** a rhododendron forest (above) in Sagarmatha National Park, Nepal, highlight a different kind of beauty from Everest's forbidding slopes. Rhododendrons thrive in the acidic soil of the Himalaya.

**THE BROWNISH RED,** rustic coat of a Himalayan tahr (opposite) blends in against dwarf rhododendrons in Nepal's Khumbu Valley. The goatlike tahrs live in herds on the wooded slopes of the Himalaya. Both males and females have backward-curving horns.

monitored my soil-erosion plots, I had the unparalleled opportunity to study the landscapes, landforms, plants, and animals along the way.

# THE LANDSCAPE TODAY

The trail contours along steep, north-facing slopes that are blanketed by a thick cloud forest of Himalayan fir, silver birch, and numerous tree and dwarf rhododendrons. *Lali guras*, the lavish bloodred flower of *Rhododendron arboreum* and national flower of Nepal, was in full bloom. Old-man's beard lichens draped the limbs, catching moisture from the early morning mists that were beginning to lift. On either side of the trail bloomed purple and yellow primroses and blue star-shaped gentians. These pioneers of spring foretold the arrival of the many hundreds of flowers to come.

### FLORA AND FAUNA

I pretended not to look at a lone musk deer, the fanged, tree-climbing resident of moist forests in Nepal and Tibet, that stood frozen beneath a lone rhododendron tree just below me. I carefully stepped over the large pile of his tiny, bullet-shaped scat that clearly announced that I was an invader of his marked territory. These creatures once were rare because poachers hunted them for pods in their bellies that are used for making perfume. Yet musk deer had become quite common in this part of the park, thanks to the enforcement of wildlife protection laws since the mid-1970s.

A small group of shaggy, goatlike Himalayan tahr browsed on the drier, shrubby south-facing slopes directly across the river. Ten years ago they posed a serious threat to many communities when local crops became their preferred food source. Now they are becoming increasingly scarce because the snow leopard, with its preference for tahr, has returned to these slopes. Scalps purported to come from the yeti, the legendary apelike creature that everyone's uncle claims to have seen, are probably made of the reddish hide of a Himalayan tahr, or so some scientists think.

An orange-barred leaf warbler and red-breasted rose finch sang in the distance, the first of many songbirds that would return to the region within the next

# "ONE DAY'S EXPOSURE TO MOUNTAINS IS BETTER THAN CART-LOADS OF BOOKS."

–JOHN MUIR, AMERICAN WILDERNESS ADVOCATE, 1896

several weeks. Gliding silently above was a lone Lammer-geier ("lamb vulture"), or Bearded Vulture, with a wingspan of ten feet.

The trail emerges from the cool forest into a high-altitude shrub land of dwarf rhododendron. The most conspicuous, the *sun-pati (Rhododendron anthopogon),* has an aromatic scent that is among the most pleasing that I have ever encountered—in small quanti-ties, that is. Porters often complain of "sunpati headaches" if their routes take them through too many miles of the pungent shrub, where the scent becomes overpowering.

## THE MOUNTAIN ABOVE

When I look up from my habitual concerns, I behold one of the grandest sights in the entire world—the black, pyramid-shaped, ice-clad, and wind-blasted summit of Mount Everest. Nepalis regard it as Sagarmatha ("mother of the uni-verse") and Tibetans revere it as Chomolungma ("holy mother"). The mountain successfully resisted all Western attempts to climb it for 32 years, beginning with the first British reconnaissance in 1921 through numerous attempts until Edmund Hillary and Tenzing Norgay's successful climb in 1953. Fifty-nine years later, in the spring of 2012, more than 200 climbers will summit on a single day.

**MOUNT EVEREST AND** the Khumbu Glacier (opposite) create an imposing landscape at the top of the world. Climbers reach Everest Base Camp by fol-lowing the Khumbu Glacier—the highest in the world—and tackle its icefall, where millions of tons of slowly moving ice tumble down a steep slope.

**PLANT LIFE THRIVES** (above) in the Everest region, including more than 100 different species of rhododendron.

**MOUNT EVEREST APPEARS** at far left above the clouds in this panoramic image (left) of the top of the world—taken from Gokyo Peak in Nepal, its summit only a short climb of two to three hours from the village of Gokyo (if you are in good shape, that is).

A MIST BLANKETS trees in Sagarmatha National Park, Nepal, like a shroud (previous pages). Only 3 percent of the park is forested, with a permanent snow and ice line at about 6,000 meters. The park's forests are under pressure as a source of firewood and lumber.

THE LAST RAYS of the setting sun bathe Mount Everest in a deceivingly warm glow (above). Everest's chiseled face beckons climbers from all over the world to test their endurance for a frigid climb at a grueling elevation.

Even so, the mountain continues to be shrouded in magic and mystery, and its ascent is considered the ultimate lifetime achievement by thousands of climbers and nonclimbers alike. The world's highest pinnacle casts its spell on all who dare approach it.

## THE HIGHEST

Why have so many been so fascinated by Mount Everest? A good part of the reason lies in one word: *highest*. At 29,035 feet (8,850 meters), Everest is the highest mountain with its origins on land. (Hawaii's 4,205-meter Mauna Kea rises over 9,000 meters if measured from its base on the ocean floor.) As the Sherpa people—the famous Tigers of the Snow, renowned for their climbing ability and loyalty to their Western climber clients—are fond of saying, "There's only one Everest." The mountain continues to draw adventure tourists, since no other mountain is its equal. As mountaineers know, many other mountains, some less than 250 meters lower than Everest, are far more difficult to climb technically, but they are relatively ignored in the media.

That makes me wonder: What would the landscapes and villages in the remote region surrounding K2 (8,611 meters) in Pakistan, or the equally remote Kangchenjunga (8,586 meters) on the India-Nepal border, look like today if they were higher than Everest?

## EVEREST'S GEOGRAPHY

The 1,500-mile-long Himalaya mountain range separates the Asian mainland from the Indian subcontinent. On a map it inscribes an arc striking WNW-ESE between Nanga Parbat and the Indus gorge in the west, to Namcha Barwa and the Brahmaputra gorge in the east. South of the range lies most of India, and from west to east the Himalaya form the northern borderlands of Pakistan, India, Nepal, and Bhutan, as well as the southern border of the People's Republic of China.

The mountains bear the marks of a geologically young and active range. Gazing at the Everest and Lhotse massifs from my vantage point that morning, I could make out curious horizontal bands of yellowish rock stretching across their exposed faces. Most people are surprised to learn that the famous Yellow Band of Everest actually began as sediments on the bottom of what was once the Tethys Ocean. Unimaginable heat and pressure over the eons caused these sediments to metamorphose into marble, shale, and gneiss and thrust them upward toward the skies.

The Himalaya are still growing, as evidenced by the frequency of earthquakes

> # "I STAND ON TOP OF THE HIGHEST MOUNTAIN ON PLANET EARTH, FOR ALL THE WOMEN OF THE WORLD."
>
> —EKATERINA IVANOVA, FIRST SOVIET WOMAN TO SUMMIT EVEREST, 1990

**THE RAGGED GEOGRAPHY** of the Everest region looks like peaks of whipped cream in this photograph taken from the International Space Station. From this vantage point looking south over the Tibetan Plateau, both Everest and Makalu are visible. It's hard to imagine that the sediments making up Everest's peak were deposited in a warm ocean millions of years ago.

in the region. While leading the Andean-Asian Glacial Lake Expedition in Khumbu in September 2011, we experienced a magnitude 6.9 earthquake that was powerful enough to make the walls of our lodge flex like rubber. Within seconds, it triggered massive, thundering avalanches that resulted in the tragic loss of life in Nepal and especially Sikkim, where the earthquake's epicenter was located. It also caused considerable geomorphic instability throughout the region. I witnessed massive rockfalls and avalanches throughout much of the next morning's hazardous trek to lower terrain, and we had to change our planned route on several occasions to a safer alternative.

## ICE AS CREATOR

Mountains in the region have been largely shaped by the sculpting action of ice. Valleys are U-shaped after thousands of years of glaciers gouging their relentless paths. Debris-covered glaciers, an indication of glacial stagnation that began with the end of the Little Ice Age in the 1850s, dominate the valleys of the high country. Lateral moraines—the deposits of boulders, rocks, and soil that form on each side of a glacier—are a common sight in the upper Imja and other valleys of the park. A series of parallel lines indicates the different periods of glacier growth and retreat. The famous Tengboche monastery is located on an ancient glacier terminal moraine, or deposit of material that indicates the farthest extent that the glacier reached. The villages of Khumjung and Kunde are located in a valley that is all that remains of a glacier that has long since disappeared. The spectacular face of the 6,812-meter Ama Dablam ("mother's necklace") was formed when several glaciers struck off in different directions from the summit region during the last ice age 12,000 years ago, eventually carving a "horn" similar to that of the Matterhorn.

More recent evidence of this dynamic environment can be seen in the devastating scouring, riverbank undercutting, and rocky depositions in the vicinity of Ama Dablam. In 1977, a glacial lake burst out in a flood that dammed the Imja River for several weeks. Torrents, or large gullies that are flooded periodically from excessive snow melt or rainfall, have sliced their way down many a mountain slope. No one who has experienced the sounds and sights of a massive avalanche will ever forget it. Smaller earthquakes or tremors are common. Less dramatic are the landscapes marked by saturated soils moving slowly downward on steep mountain slopes, high above the trekking trails. Known as solifluction (moving soil) lobes, these strange, globular landforms must be viewed either from a helicopter or high up on a neighboring mountain.

**A HIMALAYAN VALLEY**
showcases its U shape, carved by glaciers in the Tibetan Plateau. The plateau averages about 15,000 feet (4,570 meters) above sea level, with the tallest Himalayan peaks rising to nearly twice that elevation.

EVEREST

## THE THREE SIDES OF EVEREST

Most people identify Mount Everest with Khumbu, or the southern side of the mountain, a region inhabited by the charismatic Sherpa people and, more recently, Rai laborers and lodge managers from farther south. The bulk of the annual climbs, treks, research, cleanup expeditions, TV specials, IMAX and Hollywood films, and special events—such as the May 2011 summit climb and paraglide down to Syangboche airstrip by Sano Babu Sunuwar and Lhakpa Tsering Sherpa—take place on this side.

But Everest contains two other inhabited, or seasonally inhabited, sides as well. All of the early 20th-century British climbs scaled the north side, in the Tibet Autonomous Republic of China, prior to the closing of Tibet and opening of Nepal in the late 1940s and early '50s. On the north side, commonly associated with the famous Rongbuk monastery, a road used by motor vehicles leads to Base Camp, a tent village that hosts several dozen foreign climbing expeditions per year and, more recently, thousands of Chinese tourists.

The remote Kama Valley in eastern Tibet is the gateway to the technically challenging Kangshung, or East Face, of Everest, which has been successfully

A VIEW TAKEN from GeoEye's Ikonos satellite shows Everest's sharp relief in shadow and light. At this scale it seems impossible that humans climb the peak, but at least 15 different routes have been carved to the top.

THE RED SPLASH of wildflowers alongside the Kangshung Glacier adds contrast to the white snow and ice of Mount Chomo Lonzo. Its northwest and central summit were first climbed in 2005 by a French team. Its main summit was first climbed in 1954.

climbed by only a handful of people. Since its access is so difficult, this seasonally inhabited valley is a much more remote and pristine region than either of the other two sides. The British soldier, explorer, and botanist Charles Howard-Bury called it "one of the most beautiful valleys in the world" while leading a reconnaissance of the region in 1921. British soldier and explorer Sir Francis Younghusband called it "the most beautiful valley in the whole Himalaya," a view echoed by American author/conservationist Daniel Taylor, who is largely responsible for establishing the Qomolangma Nature Preserve that protects the valley.

A low-pressure system caused by prevailing winds sucks much of the moisture during the monsoon season up the Arun River and into the Kama Valley, creating conditions much wetter than those in Khumbu or the Rongbuk region of Tibet. Old-growth forests of spruce, fir, hemlock, and rhododendron ascend to altitudes of 4,572 meters. Many of the trees reach heights of 30 feet and diameters of 10 feet or more. Clouded leopard, Himalayan tahr, musk deer, Himalayan black bear, barking deer, and other large mammals are found in these forests in relative abundance. As opposed to the dry, eroded, human- and cattle-impacted alpine shrub/grasslands of Khumbu, the alpine slopes in the Kama Valley are lush and verdant grasslands, a seasonal breeding ground for yaks since the 1920s. The shrub juniper and dwarf rhododendron have been removed by yak herders to increase grazing land, but this conversion to grassland has also facilitated the dramatic growth of once rare blue sheep populations, which in turn has encouraged the return of the equally rare snow leopard during the past several decades.

Access to the valley is limited to only several months per year because snow blocks most of the passes during the others. Only 200 foreigners—their luggage carried by yaks, since portering is unknown in Tibet—visit the region per year. This daunting trek takes two weeks to complete, and any expedition must be entirely self-sufficient in terms of emergencies and evacuation plans, or otherwise face a very long trek out on the back of a yak. Located downwind of the Everest snow plume, the valley's passes can also be blocked by more than three feet of snow falling over the course of one day, making escape by foot virtually impossible. Changing weather patterns in the entire Everest region, especially in the form of more and more catastrophic rain and snowstorms, add a further element of risk to the logistics of visiting this sublime hidden valley.

## EVEREST'S CLIMATE

Geographically, the Khumbu region lies within the subtropical Asian monsoon zone, where more than 80 percent of the annual precipitation falls between June and September, although the full force of the monsoon is often mitigated by mountain ranges including the Numbur, Kongde Ri, Thamserku, and Kang Taiga Himal. For example, the average annual precipitation in Namche Bazar is 45 inches per year, but it decreases with altitude: Less than half falls on the alpine village of Dingboche, 3,500 feet higher. On the north side of Everest, the Himalaya represent a much higher topographic barrier to the monsoon, and the region is dry year-round. Partly as a result, glaciers on the north side of Everest are retreating at a faster rate than those on the south, since little moisture replenishes the snow that would ensure the continued growth of glaciers. The Kama Valley to the east is greener and moister than either the Khumbu or Rongbuk regions of Everest, since it enjoys the full force of the annual monsoon.

# "THE MOUNTAIN IS ALWAYS STRONGER THAN YOU."

—UELI STECK, SWISS MOUNTAINEER WHO SUMMITED EVEREST WITHOUT SUPPLEMENTAL OXYGEN

A DUSTING OF snow doesn't bother these yaks in Namche Bazar, a market town in Nepal. Trekkers and climbers catch a glimpse of Everest here while they spend time acclimating to the thin air. The sounds of yak trains, used to carry supplies in the region, fill the air during trekking season.

Winters are normally dry, although heavy snowfalls, sometimes catastrophic, are experienced from time to time. In November 1995, I had camped one evening on the north-facing slopes of Ama Dablam when I felt a curiously warm wind moving up the valley. The unusual red color on the summit of Everest clearly signaled to me one thing—the weather was changing, and fast. It was snowing lightly when we broke camp the next morning, but I decided to take the porters down to Tengboche, just to be safe, instead of advancing to the next work site. By the time we reached Tengboche, the snow had piled up to our knees. It soon turned heavier and continued throughout the day and night.

The next morning rumors swirled through the village, telling of hundreds of tourists, climbers, and porters in the upper valleys, huddled in huts or lodges. Ten feet of snow had fallen. More than 700 people were stranded, and tragically 26—13 Japanese and 13 Sherpas—died when an avalanche destroyed the yak herder's hut to which they had retreated in the Gokyo Valley. Thankfully, over the next few days the Nepali government managed to evacuate all of those who were stranded, a feat that I still believe is one of the most remarkable rescues in the history of mountains and mountaineering.

## A CHANGING CLIMATE

Ten years ago, the topic of climate change in the Everest region was practically unheard of. Today the topic is widely discussed among visitors and Sherpas alike. Dramatic indicators of climate change can be seen in phenomena such as the recession of glaciers and the formation of new glacial lakes. During the past three decades in the Khumbu region alone, 33 new lakes have formed and 22 existing ones have expanded. Much of the local discussion of climate change centers on the growing unpredictability of weather patterns and the impact on crops. Traditional planting dates are often delayed because of the late arrival of spring rains, and young crops are vulnerable to the often devastating downpours once the rain does arrive. People claim that more snow and more rain are falling than at any time in living memory. The monsoon used to end by the end of August but now extends well into, if not to the end of, September. Thousands of tourists can be stranded for weeks at a time at Lukla airport, a common gateway to Everest.

# THE VEGETATION OF EVEREST

As on all mountains, vegetation patterns on Mount Everest depend largely on altitude, slope, aspect, precipitation, geology, and human use. On the south-facing

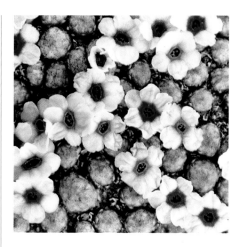

AT LOWER ELEVATIONS, rich forests of pine and birch blanket the Hinku Valley of Nepal (opposite). These forests thrive below the permanent snow line at 6,000 meters. Above the tree line, vegetation can grow all the way to the edge of glaciers.

ON THE TIBETAN side of Everest, the flowers of *Androsace tapete* add a splash of color (above). These high-alpine plants flower in July and August. Lovers of wildflowers have much to look at in the Himalaya, including orchids, rhododendrons, blue poppies, and begonias.

# "WE MUST CHOOSE THE PROPER PATH—ONE THAT DOESN'T STRADDLE TWO WORLDS, BUT RATHER UNITES THEM."

—NORBU TENZING, SON OF TENZING NORGAY AND VICE PRESIDENT, AMERICAN HIMALAYAN FOUNDATION

side, warm and dry shrub grasslands were created by herders hundreds to thousands of years ago, and they now cover the highly modified, but stable, slopes. On the moist and cool north-facing slopes, fir, birch, and rhododendron forests grow. Above 4,000 meters in altitude, shrub juniper and dwarf rhododendron also serve as the geomorphic glue that holds the thin, young, and fragile alpine soils in place. Above 5,100 meters, only sparsely distributed cushion plants can survive.

At the entrance of Sagarmatha National Park in Monjo, altitude 2,800 meters, Himalayan blue pine dominates the landscape, sprinkled with Himalayan hemlock, prickly oak, rhododendron, and Wallich's yew—trees that benefit from the heavier rainfall at these lower altitudes. Fir-birch-rhododendron forests occur between 2,500 and 3,900 meters, mostly on north-facing aspects. The uninhabited north-facing slopes of Thamserku and Kongde Ri are covered in particularly thick forests of this sort. Maple, whitebeam, and tree juniper occur occasionally, as do a great variety of dwarf and tree-form rhododendrons, particularly colorful in the early to late spring, blossoming and fading in sequence until the beginning of the monsoon rains. Huge ancient juniper trees cluster around monasteries where they have been protected for hundreds of years, showing how large these and other trees can grow if left undisturbed.

## PLANTS AND ALTITUDE

Looking at the north face of Thamserku, another peak in Khumbu, you can see the gradual upward transition from forest to tree line, and then from bands of tree rhododendron, shrub rhododendron, and dwarf rhododendron up to the alpine grassland that continues up the zone of snow, ice, and discontinuous plant cover beginning above 5,000 meters above sea level and continuing to the permanent snow line at 6,000.

Plants in the alpine zone (4,000 to 5,000 meters up) are highly specialized, containing most of their mass underground as roots with protective aboveground features such as dense rosettes or thick insulating hairs that can survive in extreme environments. In 1938, the famous mountaineer Eric Shipton found a saw-wort (*Saussurea gnaphalodes*) on a slope of scree (or loose rock debris) at an altitude of 6,400 meters on the north flank of Mount Everest—the world record for the highest known vascular plant growth.

# DISTURBING THE LANDSCAPE

Alpine ecosystems throughout the Khumbu region have been heavily affected by human activities, especially during the past 20 to 30 years with the exponential growth of mountaineering, trekking, and adventure tourism. The most common damage includes the removal of slow-growing alpine shrubs and cushion plants by lodge owners, porters, and climbing parties for use as fuel. Such clearing accelerates soil erosion and decreases slope stability, as I was able to quantify in the 1980s. Other disturbances in the alpine zone are caused by increased numbers of pack animals, primarily yak and dzo (a mix of yak and cow), brought in to accommodate growing numbers of tourists. Waterborne health hazards have accelerated as a result of improper human waste disposal. Camping sites, base camps, and high camp regions are marred by accumulations of solid waste, and landfills and human waste pits are growing in number around the villages along the main trekking routes.

In 2004, the Mountain Institute and American Alpine Club joined with local

**THE VIEW OUT** of a restaurant tent at Base Camp shows a rock-strewn field and the Khumbu Glacier (above). Some 800 people called it a temporary home during the 2012 climbing season. Base Camp is wired for the Internet and has phone service, giving climbers a connection to the wider world as they get ready to attempt the summit.

**SIGN INDICATING THE** route to Everest Base Camp (opposite)

SHERPA EMPLOYEES OF Fritz Müller, a geographer who studied the Khumbu Glacier, stand on the summit of Kala Pataar in 1956. This peak is now a popular trekking destination in Nepal on the route to Everest Base Camp.

people in the upper Imja Valley, gateway to the Everest Base Camp, to form the first Khumbu Alpine Conservation Committee. This community organization began active efforts to protect and restore its alpine ecosystems by promoting the use of imported kerosene as a fuel alternative to wood from juniper and alpine cushion plants. Juniper cover has regained a strong foothold in the last decade, but much more work remains to be done to remedy the growing problems of human and solid waste management.

Most of the thousands of people who visit Khumbu each year do so in the fall or spring seasons, avoiding the June-to-August monsoon rain and attendant leeches found at lower elevations. Having spent a summer there in the 1980s, however, I have always felt that the beauty of the Khumbu monsoon period is one of the world's best kept secrets. The days are normally sunny and partly cloudy until the early afternoon, when the clouds finally make their way up the valleys and to the villages. The rains, usually mild, arrive in the late afternoon or evening, encouraging the subalpine and alpine wildflower bloom throughout the period, beyond compare for variety, color, and density. One especially rewarding project, our pressed-plant collection, is now housed in the University of Colorado Herbarium, and it appears in the *Khumbu Plant Catalog,* compiled by the famous German botanist Dr. George Meihe for the British Museum.

# PEOPLING THE LANDSCAPE

The northern side of Mount Everest is sparsely populated. The only year-round residents are the Buddhist monks at the Rongbuk monastery, and the Kama Valley is uninhabited except for seasonal Tibetan yak herders. The 3,500 Sherpa people who live on the Nepal side are believed to be descendants of the original Sherpa pioneers who crossed the Nangpa La from Tibet 500 years ago. Their name tells the story: Sherpa means "from the east" in the Sherpa language.

## THE SHERPA PEOPLE ARRIVE

Lhakpa Norbu Sherpa, the second warden of the Sagarmatha National Park and the first Sherpa to earn a Ph.D. degree, believes that the migration was linked to political instability as well as economic conditions associated with the Little Ice Age cooling. The anthropologist Michael Oppitz, citing late 18th-century Sherpa texts, says that these migrants found uninhabited Khumbu "completely, from the highest height to the deepest valley, overgrown with thick virgin forests" and populated by an abundance of wild animals. "The rivers had no bridges, the cliffs no steps; there were no footpaths, no dwellings, no fields of grain, no woven cloth, . . . no cows to milk." These early settlers "destroyed most of the forests and transformed the landscape into agricultural fields and pastures for cattle."

Tibetan Buddhists consider Khumbu a *beyul*, a sacred hidden valley whose secrets will be revealed to the faithful when the world approaches its end. Oppitz and sacred mountain specialist Ed Bernbaum report that Pangboche and Dingboche villages were already well known as meditation sites by the time of the migration. Sherpa legends state that shepherds grazed their herds here well before the arrival of their ancestors, and geographer

THAME MONASTERY IN the Khumbu valley, 1956 (above)

TWO SHERPAS PREPARE the field to plant buckwheat in the Nepali village of Khumjung (below). The village, while not on the well-traveled path to Everest, is a destination for devout Tibetan Buddhists. Buckwheat is an important staple crop in Nepal, providing about one third of the diet in the high hills.

Stan Stevens writes of "certain ruins in high places in the Dudh Kosi valley" believed to have been early shepherd huts.

## THE STORY THE SOIL TELLS

While digging soil pits in the shrub grasslands above the village of Khumjung in the 1980s, I noticed that the soil profile was more typical of a wet forest environment rather than a dry, treeless hillside. I also found fairly frequent deposits of charcoal from burned trees at different depths, so I decided to collect charcoal and soil samples at five-centimeter intervals that could be later analyzed for pollen content, tree species, and dates of deposition. Over the years I continued to collect these samples, from the far western reaches of the park to the high alpine regions near the Everest Base Camp.

The results show that these regions were cleared at a much earlier time than first supposed. According to analyses conducted by the University of Colorado's Institute of Arctic and Alpine Research, humans began cutting and burning the south-facing Himalayan forests between 3,000 and 5,000 years ago. The fir-birch-rhododendron forest still common on most north-facing slopes today was

**MERCHANTS ENJOY A** brisk trade at Namche Bazar in Nepal. The village functions as a vital high-altitude trade stop between southern Nepal and Tibet at 11,000 feet (3,350 meters).

**Phortse, Nepal: Mani stones, tablets carved with a Buddhist devotion or mantra**

Many know the origin of the Western name for Mount Everest: When the mountain was determined by triangulation to be the highest peak in the world, the British surveyor general of India named it for his predecessor, Sir George Everest.

Far fewer Westerners, however, know the true meaning of its Tibetan name.

Almost every book on Everest mistranslates the Tibetan name as "goddess mother of the universe," based on an assumption that the Tibetan and Sherpa people who live near its base must revere the highest peak in the world as the sacred abode of a supremely important deity. But the persistent use of this translation reflects the great importance that outsiders, rather than the local Tibetans and Sherpas, place on Everest.

Everest is, in fact, a sacred peak, but a relatively minor one. Its Tibetan name—Chomolungma or, more properly, Jomolangma—comes from the name of the goddess believed to dwell there: Miyolangsangma or Jomo Miyolangsangma. The meaning of the full name of the goddess is obscure. A possible translation is "the immovable goddess mother of good bulls," which may explain why every year at the festival of Mani Rimdu at Tengboche the monks release a yak to wander freely in the mountains as an offering to the goddess.

Jomo Miyolangsangma is a relatively minor goddess. As far as I know, there is no specifically designated Goddess Mother of the World or Universe in Tibetan Buddhism, and the idea of such a goddess doesn't fit in a religion that doesn't believe in a monotheistic supreme creator. If Everest were the abode of so major a deity, Sherpas such as Tenzing Norgay would not have climbed the mountain: As devout Tibetan Buddhists, they would have regarded its summit as too sacred to desecrate.

So why isn't Everest revered as the abode of a supremely important deity? Before the British measured the height of the peak, the Tibetans and Sherpas who named it had no idea that it was the highest mountain in the world. Even if they had known, their cosmology speaks of another mountain that dwarfs Everest: Mount Meru or Sumeru, a mythic peak at the center of the universe that rises more than 80,000 miles to the heights of heaven.

Perhaps with this mountain in mind, the Tengboche Rinpoche once asked me, "How do you know Everest is the highest mountain in the world until you yourself have seen them all?"

**—EDWIN BERNBAUM Climber, speaker, and author Edwin Bernbaum directs the Sacred Mountains program of The Mountain Institute, promoting the spiritual and cultural significance of mountains.**

opened up year after year, most likely by Rai or Gurung ethnic groups from the south transforming the land for cattle grazing. Pollen analyses show that ferns became much more common starting 2,000 years ago, indicative of the drier, open woodland conditions that were being created. Birch, alder, fir, pine, and other trees and shrubs decreased starting about 1,500 years ago. Grass has dominated the landscape for 800 years now.

So what did the early Sherpa settlers see when they first laid eyes on the Khumbu region in the 1500s? The landscapes may have been similar to those found today—shrub grasslands on the drier south-facing slopes, and cloud forests on the cooler and moister north-facing slopes. All in all, not a bad place to make a home.

## THE SHERPA ECONOMY

Khumbu specialist Margaret Jeffries writes that the Sherpa economy was vitalized in the 1800s by the introduction of the potato, most likely from British colonial gardens in Darjiling. In addition, the accidental crossbreeding of the yak and cow produced an animal—the dzo—that lived longer, produced more milk, tolerated lower altitudes, and became a new trade item with Tibet. Namche Bazar became a lucrative trading crossroads where Tibetan traders brought salt to exchange

**A BRILLIANT SUNRISE** illuminates sacred Gokyo Lake in Nepal (below). Bright blue Himalayan lakes offer a spectacular backdrop for the high peaks of the Himalaya, but warming temperatures have meant that meltwater from the glaciers forms these lakes. Some are dangerously close to breaking through their moraines, or walls of glacial debris.

**THIS MAP (OPPOSITE),** published in the June 1982 issue of *National Geographic,* shows the peaks of Sagarmatha National Park. The elevation of Everest was reevaluated using a GPS receiver placed on the summit in 1998 by Wally Berg. It now measures 29,035 feet (8,850 meters).

Cho Oyu
8,188 m
26,864 ft

Mount Everest
(Sagarmatha)
8,850 m
29,035 ft

Lhotse
8,501 m
27,890 ft

Nuptse
7,861 m
25,790 ft

Traditional route to Base Camp
on Khumbu Glacier, starting
point for all approaches to
Everest from the south

Island
Peak
6,189 m
20,305 ft

Ngojumba
Glacier

Ama Dablam
6,812 m
22,349 ft

Kala Pattar
5,545 m
18,192 ft

Jasamba
Glacier

Lobujya
(herders' camp)

Pheriche

Pangboche

Khumbila
5,761 m
18,901 ft

Koner

Thyangboche
Monastery

Bhote
Kosi

Phorcha

Khumjung

Imja
Khola

Thami
Monastery

Samde

SYANGBOCHE
AIRSTRIP

Namche
Bazar

Dudh
Kosi

Orange lines define northern
boundaries of the park's
restricted-access areas

HIMALAYA

SAGARMATHA
NATIONAL PARK

CHINA
(TIBET)

HIMALAYA

Kathmandu

NEPAL

INDIA

Darjeeling

for manufactured goods from Nepal and India. By the 1900s populations were increasing, new temples and monasteries were built, and young men began a seasonal migration to Darjiling in search of employment. The Scottish explorer and climber Alexander Kellas was the first person to recognize the talents of the Sherpas as high-altitude porters, after being impressed by their strength, commitment, and good humor when he took a group on a climbing expedition to Sikkim in 1920.

The 1949 Chinese invasion of Tibet closed the border and ended trade between the two countries. Trade with Tibet would be replaced by tourism beginning in 1950 as Nepal opened to outsiders. Sherpa resourcefulness, good luck, timing, and location all have allowed them to adapt to what could have been catastrophic changes in their traditional way of life. The role they came to play in climbing expeditions, beginning with the first attempt on Everest by the Swiss in 1952 and broadening to the trekking groups in the early 1960s, would have a revolutionary impact on Sherpa culture.

## PRESERVATION CHALLENGES

In 1975, the government of Nepal gazetted Sagarmatha National Park, 430 square miles. New Zealanders expert in forestry and environmental protection worked with the staff of Nepal's Department of National Parks and Wildlife Conservation to design a management plan. This strategic and beneficial partnership drew on the New Zealanders' extensive experience in managing their own mountain parks as well as their long familiarity with the Khumbu region in general, first embodied in none other than Sir Edmund Hillary.

At first the Sherpa people did not greet the park and its staff with enthusiasm. Some muttered, "Hillary first gave us sugar, but now he's rubbing salt in our eyes." The "sugar" referred to the schools and clinics he established, but the "salt" was the national park. Many feared the park would infringe upon their way of life and restrict their use of natural resources. Reforestation exclosures built above Namche Bazar in the 1980s by park wardens Mingma Sherpa, Lhakpa N. Sherpa, and Nima Wangchu were criticized as taking away valuable grazing land. Because their browsing habits heavily damage vegetation, goats were banned and removed from the park, which angered the goat owners. The sudden presence of the military, assigned to patrol against poaching in national parklands, created new tensions. Regulations prohibiting the harvesting of shrub juniper in the alpine zone continued to be ignored by dozens of trekking lodges. Cord upon

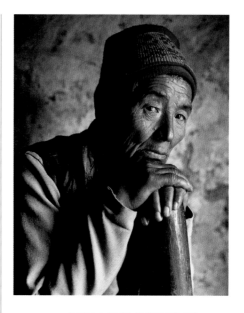

SARA LAMA (ABOVE), 70 years old in the photo, leans against a large pestle used to grind grains and rice.

TENGBOCHE MONASTERY (OPPOSITE), built in 1916, survived a 1934 earthquake and a 1989 fire.

"MOUNTAINEERS SOMETIMES PROFANE THE GODS AND OFFEND LOCAL PEOPLE... BY TRAMPLING ON SACRED SUMMITS."

—GEORGE SCHALLER, AMERICAN CONSERVATIONIST

Everyone knows that the climbing season on Everest is in the spring. Usually. I am already a bit different, since I always want to ski *down* any mountain that I climb up, so I didn't bother with the norm when starting my planning.

My theory was that the summer monsoon season would bring lots of precipitation, and I hoped that would translate to more snow on the mountain during the *post*-monsoon season, so I could ski. I also knew that the mountain had been attempted from the Nepal side only twice during the post-monsoon season in the previous six years.

The image of an Everest with few other people on it was a big draw for me.

As it turned out, we were the *only people on the mountain*.

With safety and the natural order of things in mind, our small ski team of my husband, Rob, and our close friend, Jimmy Chin, agreed on priorities before we left Base Camp: 1. Climb up. 2. Ski down. 3. Document it.

On October 18, 2006, the three of us did ski from the summit, although several members of the team, including Rob, ran out of oxygen at the Hillary Step, and so in an effort to stay out of the evening news we put our skis on our packs and down-climbed most of the way from the

**Skier Kit DesLauriers ascends the Lhotse Face (below) and skies down it (opposite).**

Hillary Step to Camp IV. After an unplanned second night at the South Col, Jimmy and Rob and I stepped into our skis outside our tents and began our descent via a direct line from Camp IV to the Western Cwm.

Usually we ski mountaineers who are still alive make a habit of climbing what we ski, but this pitch of the Lhotse Face was an exception. Climbing it would have been an undue risk on the way up the mountain that might have jeopardized the success of our first goal, which was simply to climb.

### DOWNHILL PROGRESS

As we left Camp IV at 9 a.m. on Oct. 19, I went first. After we skied for a minute on a relatively gentle 45-degree slope, the Lhotse Face became significantly steeper ahead—so steep that I couldn't see below, and the patches of snow among the blue ice that I had easily been linking together now became fewer and farther in between. I stopped, looked uphill, and when Rob and Jimmy came into sight made the X sign with my ski poles to signify that this

was a very dangerous route and maybe they shouldn't follow me. But knowing we were all still going to ski it, I next made an enormous shrug of my shoulders with my ski poles as extensions of my arms to show as if to say, "I don't have any idea how to make it any better, so you choose."

The face was over 5,000 vertical feet of an uninterrupted 55 degrees and the type of terrain with zero opportunities to stop and rest safely. Although we didn't speak about it, I was aware that if one person made one mistake, they would have been unrecognizable at the bottom. There were mandatory moments of having to leave the relative safety of a patch of firm snow and ski across frozen bare blue ice dotted with small rocks and ball-bearing-size pieces of white ice in order to reach another patch of firm snow on which to make a ski turn. Each of us remembers feeling as if we were free-soloing down a 5.13X slope (climbing grade for a route with consequences where death is a real possibility).

At times I wouldn't see Jimmy or Rob for 10, 20, or 30 minutes at a time. Each one of us had a simple light glacier

ax but no rope beyond small pieces of cord and no rescue gear beyond the standard personal crevasse rescue kit. At only one point during the almost two-hour descent did I come close enough to Rob to have a conversation. He asked, "How are you doing?" I replied, "I am scared and I don't want to die." He said, "Good. Make a plan and keep going." And we continued down.

### TEAMWORK ALL ALONE

We had studied the face, and we had watched the film *The Man Who Skied Down Everest,* so we knew that after arriving at the house-size boulder where Yuichiro Miura had fallen, we would have to move a bit to the right to cross the bergschrund at its narrowest point. The easiest crossing still required what I will call a lift of our ski tips. Not quite fully catching air, but yes, it was a small gap jump. I had to psych myself up for it in order to hit it safely with speed.

Now that I have lived it, I am not convinced of the theory that Everest is seeing more snow during the post-monsoon season, especially as our climate continues to change. While we were acclimating in September, there was such a storm that all the snow avalanched our planned route off the Lhotse Face and we had to start from scratch again. And it *is* Everest, and so then the wind blew.

I do know that the days are shorter and colder with a summit date in fall instead of spring, yet I wouldn't do anything differently next time—if there was a next time, which there won't be. Being part of a team is inherently about self-reliance and working together, so if our bond is stronger because of the season and our experience alone on the mountain, then that's the way I like it.

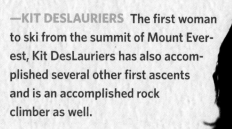

**—KIT DESLAURIERS** **The first woman to ski from the summit of Mount Everest, Kit DesLauriers has also accomplished several other first ascents and is an accomplished rock climber as well.**

**A MOUNTAINEER WALKS** in front of large ice pinnacles on the East Rongbuk Glacier on Everest's northern side.

cord of the slow-growing shrub fed the growing woodpiles, while the hillsides became increasingly denuded and eroded.

In the early 1980s I experienced the hazards of not communicating with local people firsthand. Without thinking to ask permission, I started installing my soil-erosion study plots and surveying the hillside profiles above the villages of Khumjung and Kunde. Soon my string fences, rain gauges, sediment-collection troughs, and other instruments were being destroyed during the night. Local people had seen me working and had jumped to the conclusion that I was a surveyor, intent on taking away their land. A village meeting resolved the issue, and my plots were never again touched. (I did, however, occasionally find that young yak herders had peed in my rain gauges, which I chalked up as an everyday boyish prank.)

Today many of these former attitudes have changed, largely because the park's operations have become much more inclusive, according to my colleague Ang Rita Sherpa of Kunde village. (His father, Mingma Tsering Sherpa, was Edmund Hillary's main construction supervisor, largely responsible for building the schools, airstrips, and hospitals in the region.) Local residents now view the blue pine and fir forests reestablished on the slopes above Namche Bazar with pride. Most lodge owners in the upper Imja Valley alpine zone are active members of the Khumbu Alpine Conservation Committee, which has helped ban the harvesting of shrub juniper and other plants for firewood. Communities now report feeling a greater sense of pride and ownership in the park.

## QOMOLANGMA NATURE PRESERVE

Established in 1989, the Qomolangma Nature Preserve covers an area of 13,500 square miles in the Shigatze Prefecture of the southern Tibet Autonomous Republic of China. Contiguous to the preserve are the existing Langtang National Park, Sagarmatha National Park, and Makalu-Barun National Park of northern Nepal. Compared to the Nepal side, the Tibetan terrain is generally much higher in elevation but considerably less rugged, consisting of rounded hills and flat, broad river valleys. Approximately 12,000 families live in the preserve with a total population estimated to be 68,000, engaged in agriculture, animal husbandry, or both.

Unlike the high-altitude plateau that most people associate with Tibet, the Qomolangma Nature Preserve contains a diversity of landscapes that range from subtropical, densely forested river valleys below 2,000 meters to ice-clad

**A FROZEN LANDSCAPE** of Qomolangma Nature Preserve (above), the highest-altitude preserve in the world, protects not only the high-alpine ecosystem but also the culture and history of the Tibetans. The reserve includes part of Mount Everest (Qomolangma or Chomolangma), glaciers, and mixed highland and mountain ecosystems.

**YAKS LOADED WITH** supplies wait outside Rongbuk monastery in Qomolangma Nature Preserve (left). The monastery is viewed as a sacred threshold on the path to Everest, located halfway up the 20-mile-long (32-kilometer-long) valley. Climbers who want to climb the North Face must pass through the monastery, which lies at 16,500 feet (5,030 meters).

peaks of the High Himalaya at 8,000 meters and above. It also includes the beautiful Kama Valley, gateway to the Kangshung (eastern) face of Everest. The preserve provides habitat for the rare snow leopard, kiang (wild ass), and black-necked crane as well as Tibet's only populations of the Assamese macaque. Langur monkeys, Himalayan palm civets, jungle cats, musk deer, and tahrs are also found in abundance.

An important cultural feature of the preserve is the Rongbuk (Dza Rongpu) monastery, a Tibetan Buddhist monastery in the main Dzaka Chhu River Valley draining the base of Qomolangma. The monastery, believed to be the highest in the world at 5,060 meters, is located near the Rongbuk Glacier. It became known to the outside world through reports from the first British mountaineering expeditions to Everest, passing through the river valley en route to Base Camp at 5,364 meters above sea level.

Several years ago, Chinese officials entertained the idea of closing the region to mountaineering and tourism in order to give the landscapes a chance to recover from decades of unregulated adventure tourism. In spite of its remoteness and comparative lack of visitors, garbage accumulations and the inevitable burning of fragile shrub juniper for fuel were reaching unacceptable levels. The restrictions were not enacted, however, and, for better or worse, the Base Camp region has remained open to foreign visitors.

## DECADES OF CHANGE

During the 1970s and 1980s, Khumbu was frequently cited as a case study of poor land management. Scientific and popular articles at that time proclaimed that the region was suffering from extensive deforestation as the result of a growing population, the influx of Tibetan refugees in the early 1950s, increased and unregulated tourism, and the breakdown of traditional Sherpa practices of natural resource management. One widely quoted anthropologist stated in 1975 that "forests in the vicinity of the villages have already been seriously depleted, and particularly near Namche Bazar whole hillsides which were densely forested in 1957 are now bare of tree growth, and villagers have to go further and further to collect dry firewood." Another often quoted remark was that "more deforestation [has occurred in Khumbu] during the past two decades than in the preceding 200 years."

At the same time, a much smaller, but equally vocal, contingent of people held the exact opposite view point of view. "There are more trees in Khumbu now

**SHRUB JUNIPER WOOD** (below) was the common cooking and heating fuel used for Everest's visitors, and an increasing harvest was denuding the slopes. Kerosene now substitutes for the wood from these native trees, which can live to be 100 years old and more.

**NAMCHE BAZAR, SHOWN** in photos taken in 1973 (opposite top) and 2008 (opposite bottom), has grown since the days when Sherpas built stone houses. Now lodgings are multistoried and satellite dishes dot the rooflines, making Namche the gateway to Everest and Khumbu Valley's largest town.

ONCE A GLACIER, now a lake: Imja Lake in Nepal's Sagarmatha National Park is changing, as seen in these photos taken in 1956 (above) and 2012 (opposite). That's potentially bad news for people living below the glacial lake, which is located at 16,700 feet (5,100 meters). As temperatures warm, the lake grows, potentially breaking through the moraine, or wall of glacial debris. Imja is just one of a number of glacial lakes in the region.

than there were in 1950, and I have the photographs to prove it," said Charles Houston, an American physician, high-altitude expert, member of the 1953 K2 expedition, and one of the first foreign visitors to the Everest region in 1950.

## SCHNEIDER'S RECORD

Who was right? For answers, I turned to one of the best mentors that I've ever had, and whom, incidentally, I never met: the Austrian climber-cartographer Erwin Schneider, who passed away in the late 1980s. Between 1955 and 1963, Schneider completed a field survey of the Khumbu region that included the use of terrestrial photogrammetry—determining geometric properties of objects and making maps from photographs taken at high-altitude vantage points. His camera was huge and heavy, and it is now housed at the Austrian Alpine Club in Innsbruck. That he was personally carrying this camera and its equally heavy tripod in his late 50s, often scaling semi-technical mountains and outcrops above 5,500 meters to get a particular landscape photograph, came as no surprise once I learned that Schneider had been one of Austria's top alpinists in the 1930s.

In 1984, while living in Khumbu, the Swiss geologist Daniel Vuichard gave me a packet of Schneider's photographs taken in the 1950s. I instantly realized that they provided a valuable window into the past and might help to resolve

some of the controversy concerning the changes in the landscape since that time. Little did I know then that I would be retracing Schneider's footsteps, replicating his photographs of villages, mountain landscapes, and glaciers, for the next 30 years.

In my spare time in 1984, and then again in the course of research expeditions launched in 1995, 1999, 2001, and 2007, I replicated dozens of Schneider's photographs. In that way I was able to add my own research results to the landscape-degradation discussion. As early as 1984, it became clear that Charles Houston was right, after all: The hillsides above Namche Bazar had not been densely forested earlier in the century. The landscapes of the 1950s were essentially unchanged 30 years later. During the 1990s, the repeat photographs started to reveal a very encouraging phenomenon. The tree cover surrounding the main Sherpa villages, and along the Imja River Valley, was in fact increasing.

## REPEAT PHOTOGRAPHY

Replicating the photographs taken by Schneider, Houston, and others in the 1950s allowed me to track the tremendous growth and development of the Sherpa villages located along the main Everest Base Camp trail, including an almost complete change in building materials and styles. Retaking the photos also provided

> ## "I OWE A SUPREME DEBT OF GRATITUDE TO THE MOUNTAINS."
>
> —ELIZABETH HAWKINS-WHITSHED, EARLY BRITISH FEMALE MOUNTAINEER

a unique opportunity to demonstrate just how dynamic a landscape the Everest region was. The scars and impacts of destructive glacial-lake-outburst floods (GLOFs), landslides, and torrents that had occurred in the interim could be easily seen when comparing the old landscape photographs with the new.

But perhaps most important were the changes recorded in the high-altitude alpine landscapes. For years I had been perplexed by the fact that these environments produced by far the highest rates of soil loss. But with repeat photography in combination with detailed alpine groundcover studies and surveys, I discovered that more than half the shrub juniper cover of the upper Imja Valley had been lost since the 1950s, likely because of the growing use of the shrub for fuel by new lodges and more trekking and climbing groups, coupled with the lack of enforcement of park regulations in these more remote regions.

**NASA ASTRONAUTS TOOK** this photo of the northern approach to Everest from the International Space Station in January 2011. Climbers starting in Tibet (China) travel the East Rongbuk Glacier, at lower left, and then the razor-sharp North Col to the summit, just beyond the top edge of the frame.

# CHANGING GLACIERS

The then-and-now photos had other valuable uses. In the fall of 2007, I was climbing at 5,500 meters up a steep, precarious boulder field in the upper reaches of the Imja Khola watershed. The steep and extensive ridge that makes up the south faces of Lhotse, a neighbor peak to Everest, rose for thousands of meters above me across the valley. The summit of Everest was hidden because of my relatively low altitude. A light snow that had fallen the night before was turning what would have normally been a scramble into a semi-technical and quite dangerous climb. Finally, I realized that it would be foolhardy to proceed under these conditions. I called out to Pema Tema Sherpa, my research assistant and two-time Everest summiter, to stop. Our altitude, 100 meters below the cairn that Erwin Schneider had constructed to mark the location of his photo, was good enough for the objectives at hand.

Comparing the old photograph and the view in front of me now revealed a dramatic change in just 60 years. The Imja Glacier that I had hoped to rephotograph was gone, replaced by a lake almost half a mile long. Icebergs as big as a house had broken off the former glacier and now floated aimlessly in the water. The glacier had melted, leaving behind a large and potentially dangerous lake in its wake.

### MÜLLER'S RECORD

The previous year my mentor, Jack Ives, had given me a box of old photographs collected by his friend and colleague Fritz Müller. Müller, a Swiss-Canadian glaciologist, was the scientific team leader of the 1956 Swiss Everest expedition. When the climbers went home, he stayed on for nine more months, living above 5,000 meters (a high-altitude record at the time) and studying the glaciers with his Sherpa assistants. Although exceptionally strong, Müller had died of a heart attack in 1980. In the confusion following his early death, most of his photographs were lost, but one box was saved, full of 35mm contact sheets only. Over time I was to discover that in addition to notes, rock rubbings, pressed plants, letters, and telegrams, the box also contained photographs of two other noted climber-scientists, Sir Charles Evans and my personal hero, Erwin Schneider. Evans was the deputy director of the 1953 British expedition that successfully put the first humans on the summit of Everest, Edmund Hillary and Tenzing Norgay, and were it not for a failed oxygen system, he might have arrived there first.

Since then, a number of glacier inventories have been conducted, largely accomplished by using satellite images as the primary data source. In 2007, very

**GLACIOLOGIST FRITZ MÜLLER** surveyed and photographed the glaciers around Everest in 1956, providing key data points for the study of changes in the ice and climate today. Here Müller and assistants view Everest (the black slope to the left) from an altitude above 18,000 feet (5,500 meters).

# "WE LOOKED AROUND IN WONDER. WE HAD REACHED THE TOP OF THE WORLD."
### –SIR EDMUND HILLARY, 1953

few on-the-ground field studies had been conducted. This struck me as odd, given all the media statements about climate change in the Everest region. None of these studies, as it turned out, was based on actual fieldwork or in collaboration with local people. It thus became even more important to retrace the footsteps of Evans, Müller, and Schneider, climbing up and over the glaciers to their photo points and documenting what had changed.

## RETRACING FOOTSTEPS

Finding the places where the original photos were taken was the first challenge. The global positioning system (GPS) did not exist when Müller, Schneider, and Evans took their photographs, and no records of their geographical coordinates remained. But we were usually able to find most of the photo points through a combination of our own familiarity with the landscapes, the suggestions of climbing guides and yak herders, and luck (although where some were taken mystifies me to this day). I also decided that the golden rule for finding the exact locations, especially for Erwin Schneider, was to always go higher than I expected. Nothing was "good enough" for these early scientists, and they always went for the best views regardless of the difficulties involved in getting there. Several proved impossible to reach because of climate-related changes that had occurred in the interim. In some cases massive ice avalanches blocked the original trail. Certain glaciers that had been solid and healthy in Müller's day were now melting rapidly, making them hazardous to climb upon. One or two other opportunities had to be abandoned because, if truth be known, Erwin Schneider was simply a stronger climber than I am.

Receding glaciers and new glacial lakes were not the only changes I studied. The Imja Glacier, we determined, had melted so rapidly because it had

been covered by a thin layer—half a meter or so—of soil and rocks (a geologist would call it a "debris-covered glacier"). The debris would heat up during the day and transfer the heat directly to the ice below. Glaciers become debris-covered when their forward movement stagnates, occurring in the Everest region since the end of the Little Ice Age in the late 1800s. Over time, boulders, rocks, soil, and other debris from the valley and glacial moraines cascade down and cover the ice. A very thick cover of debris, with boulders up to the size of small houses, tends to insulate them from the sun's heat. But the glaciers we observed had ablated, or lost mass. An abundance of new, small, meltwater ponds now appear on their surfaces, and any exposed glacier ice has melted away.

**MÜLLER'S RESEARCH PROVIDED** numerous fascinating photographs of Mount Everest in 1956. Sherpa assistants climb a glacier, possibly near Everest Base Camp's location today (opposite and above); despite rough research camp conditions, Fritz Müller carried an umbrella against the frequent autumn snow showers.

**PHOTOGRAPHS OF THE** Khumbu glacier, as captured by Müller's camera in 1956 (below), provide evidence of shrinking ice and receding glaciers in the Himalaya today.

Dozens of smaller glaciers below 5,200 meters were gone altogether. Their low altitude and lack of protective debris cover apparently made them more susceptible to the region's warming trends. Ice above 7,000 meters, however, appeared to be remaining relatively stable, since temperatures there remain below freezing for most of the year. Such high-elevation ice and glaciers with north-facing aspects appear to have changed little during the past 60 years. Warmer, south-facing regions tended to show more signs of melting and loss of ice, and yet glaciers on the north side of Everest are melting faster, in part because the north side receives far less annual precipitation.

In summary, I discovered firsthand that climate change is happening, even in the world's highest mountains. Glaciers are melting, new lakes are forming, snow lines are receding to higher altitudes, and weather patterns are becoming unpredictable.

## GLACIAL-LAKE-OUTBURST FLOODS (GLOFS)

Concerns have been raised about the impact of these changes on future water supplies in the Hindu Kush-Himalaya region, as well as the increased likelihood of glacial-lake-outburst floods. As with Imja, dozens of new glacier lakes holding hundreds of millions of gallons of water have been created in the Khumbu region since the early 1960s. Usually contained by dams of loose boulders and soil, these lakes present a risk of glacial-lake-outburst floods, triggered by swift lake-area expansion, seepage, changes in the lake's water level, and surge waves caused by rockfalls, rock slides, or ice calving. GLOFs unleash stored lake water, often causing enormous devastation downstream that can include high death tolls as well as the destruction of valuable farmland and costly infrastructure such as hydroelectric facilities, roads, and bridges. The 1985 Langmoche outburst in Sagarmatha National Park, for example, destroyed the two-million-dollar Thami hydroelectric facility that had just been completed, hundreds of acres of cropland, and dozens of bridges downstream. The 1998 outburst of the Tama Pokhari in the Hinku Valley of Makalu-Barun National Park in Nepal destroyed trails and seasonal settlements for more than 50 miles downstream, and the damage is still visible in satellite images taken a decade later.

**SCHOOLCHILDREN WORK ON** their lessons in Khumjung School in Sagarmatha National Park. The school, built in 1961 with funding from Sir Edmund Hillary's Himalayan Trust, has more than 350 students. The trust has built more than 25 schools in the region and offered assistance to others. The Nepali government has administered all area schools since 1972.

# A CHANGING CULTURAL LANDSCAPE

As the Greek philosopher Heraclitus stated, "Nothing endures but change," and the Khumbu region of Nepal is no exception. Largely fueled by the phenomenal growth of tourism, the solitude of the Khumbu region has been altered in ways unimaginable to most Sherpa people 30 years ago. Perhaps most striking are the changes in both the number and design of houses, lodges, and shops. Between 1950 and 1973, Namche Bazar was a sleepy little village of traditional Sherpa stone houses with split fir shingles for roofs. Now it is wall-to-wall, two- to three-story candy-colored lodges built specifically for tourists. Each year they ascend higher up the slopes of the bowl-shaped locality, and one awakens and goes to sleep at night to the chink-chink-chink sound of stonecutters squaring off building stone for yet another lodge. The same is true for nearly every village located on the main trekking trail to the Everest Base Camp, including Kunde, Khumjung, Tengboche, Deboche, Pangboche, Pheriche, and beyond.

Along with these changes have also come positive impacts. With increased incomes comes improved health. Goiters, caused by a lack of iodine, were a common sight through the early 1970s, but they have now been virtually eliminated. Sherpa children receive a better education, and there are now many Sherpa Ph.D.'s, doctors, dentists, conservationists, and government workers. Most villages have electric power from mini-hydro stations, complemented with alternative technologies such as solar energy–driven water heaters, lights, and battery chargers. These devices have become particularly popular in lodges in the more remote alpine villages that do not have electric power, such as Pheriche, Dingboche, and Chukung. Dingboche even has solar-powered Internet access. Lodges in the alpine zone switched to kerosene for cooking fuel in 2004, which has resulted in the rapid restoration of fragile alpine ecosystems (a project cofinanced by the National Geographic Society in 2005 and 2007). Thanks to Sir Edmund Hillary and the Himalayan Trust, the region has schools, clinics, and airstrips—Hillary's way of thanking the Sherpa people for their help in achieving "the most meaningful of my accomplishments," as he put it.

But with the positive also come the negative impacts. With no zoning

A SHERPA CHILD calls Lukla, Nepal, home—a town that is the starting point for many Everest climbers. Located at 9,200 feet (2,800 meters), the town has a tiny airstrip fielding trekkers from all over the world.

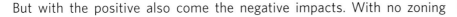

## VOICES  WATCHING CHANGE IN KHUMBU

My name is Sherap Jangbu. I was born in Namche Bazar, gateway to the Chomolungma, in 1954, exactly one year after the first ascent of Chomolungma—Mount Everest—by Tenzing Norgay Sherpa, from Nepal, and Sir Edmund Hillary, from New Zealand.

I am 58 years old and married, with a lovely wife, Lhakpa Doma. We have one grown-up son and one daughter, and we also are lucky to have one grandson, Tsering Wongchu, from our son, Mingma, and another, Tsering Paljor, from our daughter, Rita.

Namche Bazar is the main trading point in Khumbu region. All those who go on expeditions to Everest, trekking in the area, or to Tibet for trade will go through Namche Bazar, because it is suggested that you first spend two nights here to acclimate. In Namche Bazar we have everything from luxury hotels

to lodges, to accommodate all varieties of tourist.

As far as I remember, the first time I ever saw the Mik Karu (White Eye, which is our nickname for white people or foreigners) was 1963, the date of the first successful American Everest expedition. My uncle, Yula Tsering, was on that team.

At that time, only a few expeditions came to Chomolungma, and it used to take at least four to five months to finish an expedition or at least two months to finish the trekking, because they had to walk from Kathmandu to Everest and back. In 1964 Sir Edmund Hillary built a small airstrip in Lukla. That shortened the trekking and expedition time, so from the 1970s on, the tourist flow increased.

—SHERAP JANGBU SHERPA A lifelong resident of Khumbu, Sherap Jangbu Sherpa organizes and leads treks throughout the Everest region.

**DZOS SHARE THE** trail in the rhododendron forest near Teng-boche monastery (opposite). A cross between a cow and a yak, dzos, along with yaks, carry supplies and gear for local inhabitants and trekkers.

**A GARBAGE INCINERATOR** at Namche Bazar helps to handle some of the detritus from the influx of trekkers and climbers to the region. An estimated 55 tons of waste is created every year along the route from Lobujya to Everest Base Camp.

system, the number of lodges increases each season. Each year lodge owners bring in canned and bottled goods by the ton. These containers end up in landfills outside every village, since there is no recycling system or incentives for them to do otherwise. Five tons of solid waste is generated each year in the Everest Base Camp alone. Human waste and raw sewage is channeled directly into nearby freshwater steams, polluting water supplies for thousands of people living downstream. Trekkers increasingly complain about intestinal disorders contracted along the Base Camp trail in the upper Khumbu Valley, likely caused by improper sewage disposal. Porters use large blue plastic barrels to carry human waste by the ton to a place about two hours' walk south from Base Camp. There it is dumped into large pits, some of them located in seasonal riverbeds.

PRAYER FLAGS CREATE a rainbow of color at the top of Cho La, Nepal. On the route to Everest, the pass, located at 17,780 feet (5,420 meters), offers stunning views of the Himalaya.

# THE FUTURE EVEREST LANDSCAPE

What does the future hold for the world's highest mountain, with its mélange of positive and negative changes? Tourism to the Kama Valley will probably increase as a new generation of wealthy adventure tourists seeks new and more pristine paths to the mountain, steering clear of ever more trekking tourists, yaks, dzos, and porters on the Khumbu Everest Base Camp trail. The north side of the mountain could indeed be closed, as Chinese officials have threatened, giving the fragile landscape a chance to heal and technicians the time to clean up decades of accumulated garbage, tents, oxygen bottles, and, sad to say, corpses.

The fate of the Khumbu Valley is more complex. It will be largely determined by future global economies, the political stability of Nepal, and decisions made by the Sherpa communities regarding the current processes of change. Oil embargoes, the price of fuel and travel, and recessions and depressions will clearly result in ebbs and flows of tourist numbers. Some outdoor magazines advise readers to get to Nepal sooner rather than later, since political instability could once again result in the long-term closing of the country's borders to foreigners.

But world economies and national politics aside, how long will Khumbu remain a desired place to visit, given the increasingly overcrowded and unsafe health conditions? Will solid and human waste deposits in villages and along the trail reach a level that makes it an undesirable place to visit?

The short answer? Not likely. "There's only one Everest," as many have said, and that fact will probably draw future generations of climbers, trekkers, scientists, filmmakers, journalists, mystics, academics, teachers, development specialists, and thrill seekers for years to come. The Sherpa people themselves face questions that are unprecedented in their nearly 400-year history in the region. The loss of their traditional culture is matched by their decrease in numbers as more and more move to Kathmandu or New York City. They have to settle long-term management issues concerning the tourism from which a majority profit in one way or another. But if the Sherpa people have demonstrated any one attribute century after century, it's their ability to adapt, to recognize new opportunities as they arise, and to find solutions to what might appear to be insurmountable challenges.

And the mountain itself? It, too, shows a remarkable ability to adapt. Everest has survived the spread of humankind over the Earth, the sweep of ice ages and interglacial periods, the slow migration of flora and fauna up and down its slopes, and the recent discovery of its lower slopes by farmers, yak herders, and, in the past 60 years, adventure tourists. Mount Everest will continue to tower among the clouds long, long beyond human memory.

# "WE WEEP."

-BARRY BISHOP, TEAM MEMBER OF THE 1963 AMERICAN MOUNT EVEREST EXPEDITION

TWO SHERPA WOMEN differ in age but offer a study in longevity. Most of the Sherpa in the region don't work in tourism, but those who do bring important money into the economy of this poor country. The strong Sherpa spirit endures in this extreme land.

# CHAPTER 5

# THE CLIMBERS

Bernadette McDonald

Rob DesLauriers, Kit DesLauriers, Mingma Sherpa, and Dave Hahn move cautiously toward Everest's Lhotse Face.

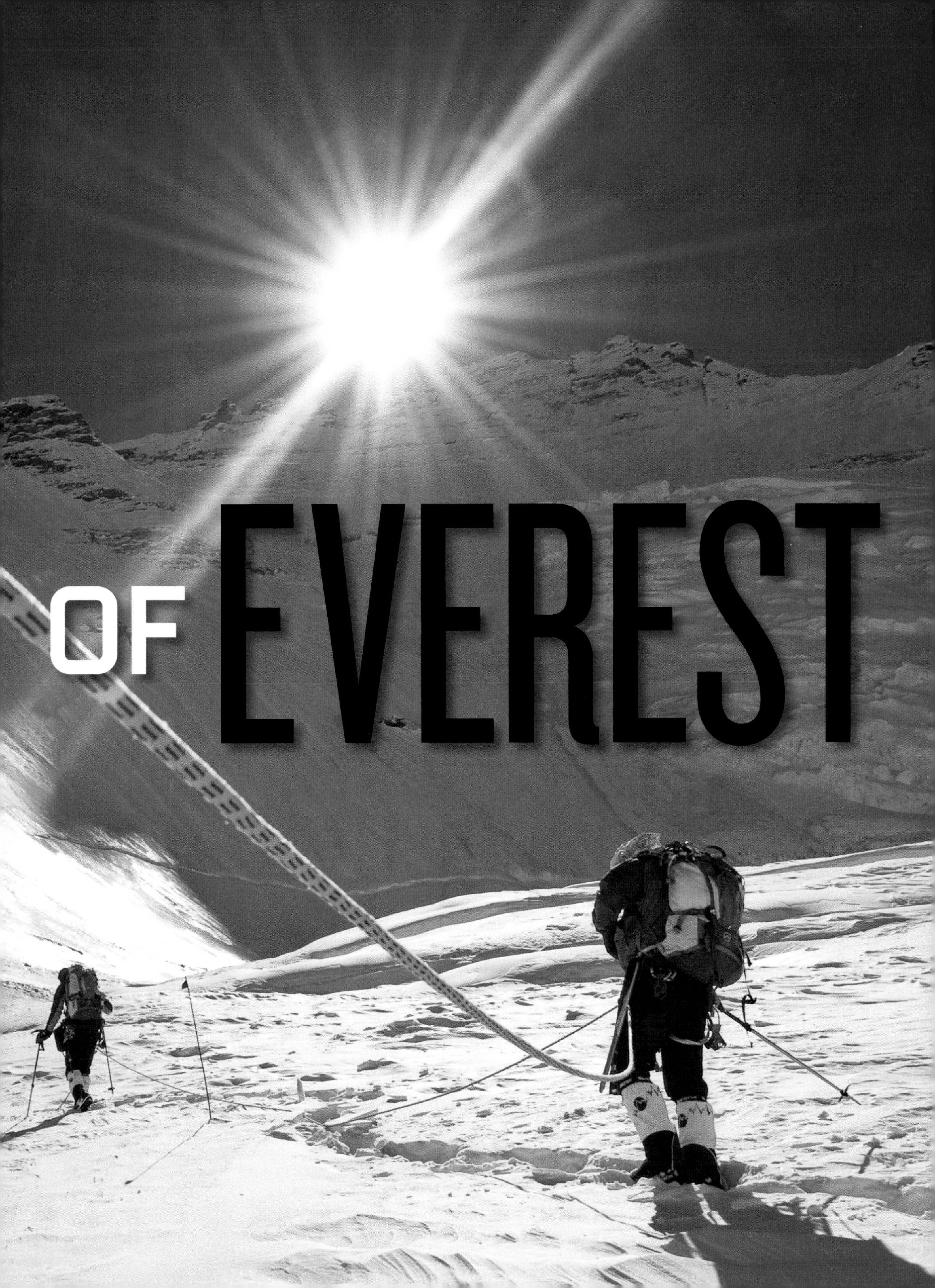
OF EVEREST

WHEN TENZING NORGAY HEAVED HIMSELF ONTO THE TOP OF THE HILLARY STEP, "LIKE A GIANT FISH," ACCORDING TO EDMUND HILLARY, THEY BOTH KNEW THEY WERE ABOUT TO REACH THE SUMMIT OF THE HIGHEST MOUNTAIN ON EARTH.

When Tenzing Norgay heaved himself onto the top of the Hillary Step, "like a giant fish," according to Edmund Hillary, they both knew they were about to reach the summit of the highest mountain on Earth. As they embraced on top on that historic day—May 29, 1953—their names became forever linked with Everest.

Ten years later, as Willi Unsoeld and Thomas Hornbein stood on top of Everest after having climbed up a couloir that would bear his name, Hornbein pondered the meaning of it all: "It is strange how, when a dream is fulfilled, there is little left but doubt."

On May 8, 1979, Reinhold Messner admitted that when he and Peter Habeler stood atop Everest, having become the first to climb it without supplemental oxygen, he felt nothing special—just a sense of calm. Only when Habeler joined him on the summit did emotion overcome them both.

Less than a year later, on February 17, 1980, in the face of screaming winter winds, Polish climbers Krzysztof Wielicki and Leszek Cichy reached the top, becoming the first to climb Everest—or any 8,000-meter peak—in winter. All the elements had been stacked against them, but they knew they had to perform. As Wielicki explained: "Poland was Poland, and Everest was Everest."

**BARRY BISHOP HAS** nothing but air around him as he rests on the summit in 1963 (above), the year he and three other Americans became the first from their country to reach the top. Even then technology for mountaineering had advanced far beyond the goggles and altimeter (below) used by George Mallory 40 years earlier.

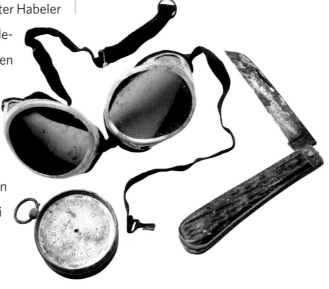

PORTRAITS OF LEGENDARY
British climbers George Leigh
Mallory (above, at left) and
Andrew Irvine fix these men in
time. They pioneered a route on
the Northeast Ridge from Tibet
but disappeared into the clouds
during their 1924 summit attempt.
It was Mallory's third time on
Everest in four years.

A BRITISH RECONNAISSANCE
team (opposite) descends from
Kartse at the head of the Kama
Valley in Tibet while searching for
a northern route to the summit in
1921. The team, including George
Mallory, appealed to the Dalai
Lama to grant them permission to
visit Tibet.

British climber Alison Hargreaves wept and laughed simultaneously as she radioed from the summit of Everest in 1995: "Tell my children I'm on the summit of the world and I love them dearly." After becoming the second woman to climb the mountain without supplemental oxygen and the first to solo the peak, Alison's first thoughts were for her children.

Outstanding moments. Amazing achievements. History being made. But what led to these remarkable exploits? They weren't accidental. Years of effort, decades of exploration, dozens of unfulfilled dreams and aspirations led to these summit moments. And it all began with a surprise measurement.

# EXPLORATORY YEARS

"Sir, I have discovered the highest mountain in the world," exclaimed Radhanath Sikdar as he rushed into the office of Sir Andrew Waugh, the surveyor general of India. His announcement shaped the history of Himalayan climbing. Until then, Kangchenjunga had held the top spot. But once the Great Trigonometrical Survey of India, under the leadership of Sir George Everest, was completed, the mountain known as Peak XV appeared higher, at around 29,000 feet (8,839 meters). Later measurements edged the total up a bit to 29,028 feet (8,848 meters).

Sir George's successor thought it appropriate to name the mountain after this man who had initiated the survey in the first place. Everest's name would never be forgotten, even though Chomolungma, which, according to Tenzing Norgay, means "the mountain so high no bird can fly over it"—is a much more descriptive title.

The Alpine Club, started in London in 1857, soon became the incubator for forays into the greater ranges. One of its early explorers, Charles Bruce, was the first to propose climbing Everest. World War I put a stop to any efforts, but once the war ended, interest began anew.

In 1919 Capt. J. B. L. Noel, a photographer and filmmaker who had grown up in exotic locales from Switzerland to India and had barely escaped World War I alive, spoke to the Royal Geographical Society, claiming to have been within 40 miles of Everest in 1913, traveling in disguise. A year later Alexander Kellas, a chemistry professor interested in the effects of altitude on the human body,

# "YOU CAN CHANGE YOUR HOBBIES, NOT YOUR PASSION."

–KRZYSZTOF WIELICKI, POLISH ALPINIST AND FIRST TO CLIMB EVEREST IN WINTER

# "MOUNTAINEERING IS A RELENTLESS PURSUIT."

—HERMANN BUHL, AUSTRIAN ALPINIST

displayed photos he had taken from within ten miles of the glaciers on Everest. Kellas and Noel were soon hatching secret climbing plans.

After obtaining permission from the Dalai Lama to mount an expedition to Everest, the Alpine Club joined forces with the Royal Geographical Society. Their strategy: a reconnaissance in 1921 and a serious attempt the following year. The reconnaissance team of four included a young climber, George Leigh Mallory. A vicar's son, Mallory was one of Britain's finest mountaineers and almost impossibly handsome.

Mallory didn't accept the offer immediately. He had a profession, and he was a married man with children. But he understood that climbing Everest could be advantageous for his career, so he accepted. The decision would make him famous, but it would also cost him his life.

In May 1921, because Nepal was closed to foreigners, the team left Darjiling and headed north through Sikkim into Tibet, accompanied by Sherpas who carried their heavy loads. Less than a month later, Everest claimed its first victim. Still approaching but within view of the mountain, Kellas died, apparently from heart failure. This was a tragic blow, for Kellas was an experienced Asian explorer and had already climbed nine peaks over 20,000 feet. He had also predicted that Everest could be climbed without the assistance of supplementary oxygen. He was right. But more than 50 years would pass before anyone did it.

As the explorers emerged into the Rongbuk Valley, the stupendous mountain reared up in front of them. Mallory described it as "a single gesture of magnificence to be lord of all, vast in unchallenged and isolated supremacy." More practically, what he saw was the 10,000-foot North Face, flanked on the left by the Northeast Ridge and on the right by the long West Ridge. An additional buttress descended in a northerly direction, completing the geometry.

For days they explored the terrain, navigating the Rongbuk Glacier, climbing small peaks in order to get better views, trying to piece together the topography, always in search of easy ground. But they missed a nondescript little stream that emerged from a narrow valley in the east onto the Rongbuk, a stream that would eventually provide the key to climbing the mountain. From a high col they called the Lho La, they could see the Khumbu Glacier and the upper basin, which they named the Western Cwm. After two months of searching, they found a way over the Lhakpa La and down onto the East Rongbuk Glacier. Not until September 24 did they finally reach the North Col, four months after starting out! Mallory was weary of the effort, vowing in a letter to his wife, "I wouldn't go again next year, as the saying is, for all the gold in Arabia." But three months later, he was on his way back.

THE SECOND BRITISH expedition tackled Everest in 1922 with a large team of porters, some of whom helped at Base Camp while a few of the most talented assisted at the highest elevations. Two team members, Howard Somervell and Edward Norton, would reach 28,128 feet (8,573 meters) without using supplemental oxygen. George Mallory and Andrew Irvine would then set out for the summit. The glove above was found with Mallory's body.

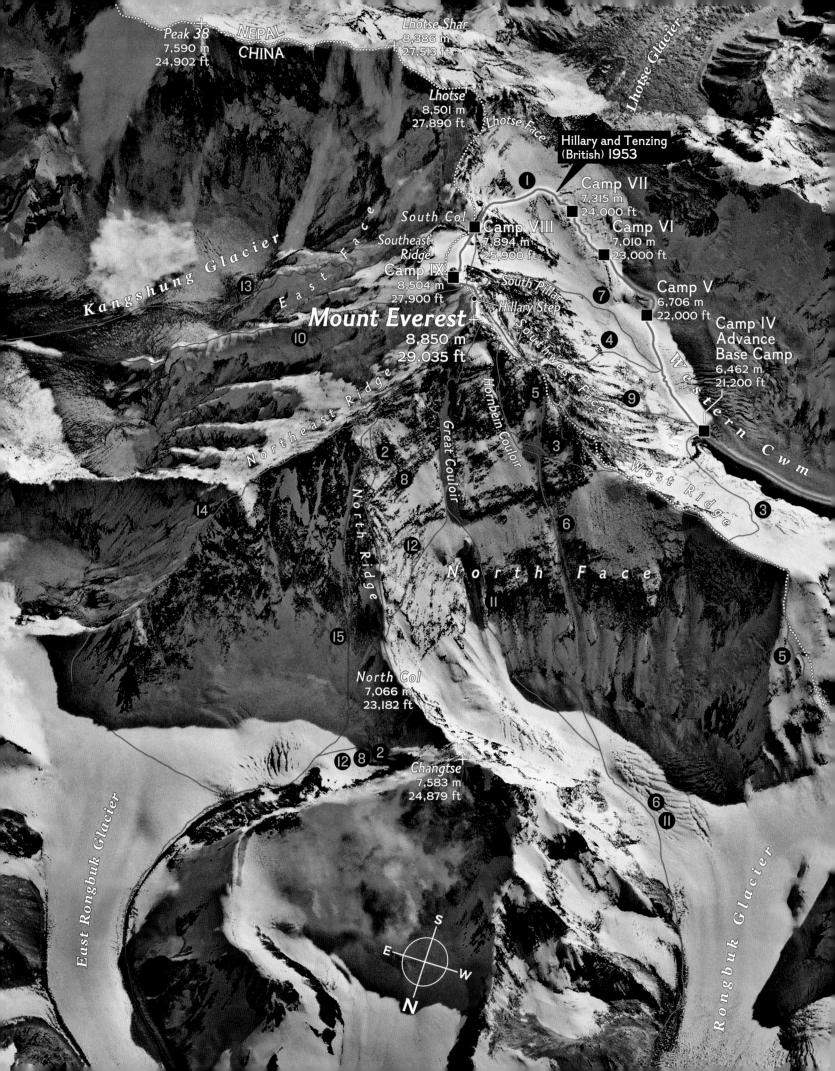

Peak 38
7,590 m
24,902 ft

NEPAL
CHINA

Lhotse Shar
8,386 m
27,513 ft

*Lhotse Glacier*

*Lhotse*
8,501 m
27,890 ft

*Lhotse Face*

Hillary and Tenzing
(British) 1953

① Camp VII
7,315 m
24,000 ft

*South Col*

Camp VIII
7,894 m
25,900 ft

*Southeast Ridge*

Camp VI
7,010 m
23,000 ft

Camp IX
8,504 m
27,900 ft

*South Pillar*

⑬

*East Face*

*Hillary Step*

⑦

Camp V
6,706 m
22,000 ft

**Mount Everest**

8,850 m
29,035 ft

⑩

*Kangshung Glacier*

④

Camp IV
Advance
Base Camp
6,462 m
21,200 ft

⑨

⑤

*Hornbein Couloir*

*Southwest Face*

*Western Cwm*

② *Northeast Ridge*

⑧

*Great Couloir*

③

⑭

*North Ridge*

⑫

⑥

③

N o r t h   F a c e

*West Ridge*

⑤

⑪

⑮

⑤

North Col
7,066 m
23,182 ft

⑥

*East Rongbuk Glacier*

⑫ ⑧ ②

Changtse
7,583 m
24,879 ft

⑪

*Rongbuk Glacier*

S

E ⊕ W

N

# 15 WAYS TO THE TOP

**1 1953 BRITISH EXPEDITION**
Edmund Hillary and Tenzing Norgay, via the Western Cwm and the South Col, May 29.

**2 1960 CHINESE EXPEDITION**
Via the North Col and the Northeast Ridge, May 25.

**3 1963 U.S. EXPEDITION**
Via the West Ridge with traverse of the North Face. Final 800 feet on the West Ridge, May 22.

**4 1975 BRITISH EXPEDITION**
Via the Southwest Face, September 24.

**5 1979 YUGOSLAVIAN EXPEDITION**
Via the entire West Ridge, May 13.

**6 1980 JAPANESE EXPEDITION**
Via the North Face, May 10.

**7 1980 POLISH EXPEDITION**
Via the east side of the South Pillar, May 19.

**8 1980 MESSNER SOLO EXPEDITION**
Reinhold Messner, via the North Col and North Face. First solo ascent without supplemental oxygen, August 20.

**9 1982 SOVIET EXPEDITION**
Via the Southwest Face and the West Ridge, May 4.

**10 1983 U.S. EXPEDITION**
Via the East (Kangshung) Face, October 8.

**11 1984 AUSTRALIAN EXPEDITION**
Via the entire Great Couloir of the North Face, October 3.

**12 1984 U.S. EXPEDITION**
Via the East Face, October 8.

**13 1988 INTERNATIONAL EXPEDITION**
Via the southern buttress of the East Face, May 12.

**14 1995 JAPANESE EXPEDITION**
Via the Northeast Ridge from the East Rongbuk Glacier, May 11.

**15 1996 RUSSIAN EXPEDITION**
Via the Northeast Ridge from a couloir to its east, May 20.

**DURING THE 1922** British expedition, George Mallory and Edward Norton climb toward their high point of 26,985 feet (8,225 meters) on the mountain (right). A couloir on the North Face would be named for Norton, now also known as the Great Couloir. Reinhold Messner would use it for his 1980 summit—the first solo ascent without supplemental oxygen.

**GEORGE MALLORY'S WATCH** and a tin of meat lozenges were found on his body in 1999 (opposite). The hands of the watch had fallen off, but rust stains seem to mark a time of 1:25. The tin of meat lozenges promised "portable nourishment at all times."

The high point of the 1922 pre-monsoon attempt came when Geoffrey Bruce, together with George Finch, Britain's best snow and ice specialist, reached 8,321 meters, both using supplemental oxygen. Amazingly, this was Bruce's first mountain ascent. But the expedition ended in tragedy when a massive avalanche caught the team on the slopes of the North Col and killed seven Sherpas.

A newcomer joined the team in 1924—Andrew Comyn Irvine, known by all as Sandy. A superb athlete with powerful shoulders, he rowed for Oxford and had first-rate practical skills and a great attitude. As his great-niece, Julie Summers, remembered, "Sandy Irvine was born to be brave." Early on, Mallory singled out Sandy as his preferred climbing partner, in part because of his physical abilities but more importantly because of his handiness with the oxygen apparatus.

Mallory had a strategy for Everest in 1924: two simultaneous attacks, one with oxygen and one without. He intended to use "gas" and make the ascent with Irvine. On this, his third trip to Everest, Mallory was the most motivated of the bunch, almost grim. In a letter to his wife he wrote, "It is almost unthinkable with this plan that I shan't get to the top; I can't see myself coming down defeated." To a Cambridge friend, he was even more blunt: "This is going to be more like war than mountaineering. I don't expect to come back."

After weeks on the mountain, expedition leader E. F. Norton put in a remarkable effort. Without supplemental oxygen and seeing double, he turned back at 8,570 meters, just 900 feet below the summit. His oxygen-free record would last for 54 years. Four days later, Mallory and Irvine, using gas, set off from Camp VI at 26,800 feet (8,169 meters). Hours later, Noel Odell was scanning the slopes when the clouds parted to reveal a small object advancing upward quickly, approaching a rock step. A second object followed. They appeared to be moving. Then the clouds moved in.

Odell was confused. The pair had reached the top of the Second Step—a steep rock section of the ridge—but they were dangerously behind schedule. Concerned, he climbed even higher than Camp VI, calling and yodeling, trying to lure the climbers down. He descended to the North Col, and the next day he climbed farther beyond Camp VI, calling to no reply. Finally he dragged two sleeping bags from the tent and placed them in a T position, a signal for the teammates below that all was lost. Odell was convinced the two had reached the summit, based on his last sighting and his ardent desire that they had. For Ruth Mallory, it didn't matter. Everest had claimed the life of her husband and the father of her children.

No other lost explorer in mountaineering history has generated more speculation than George Mallory. In 1933 an ice ax was found on a rock slab 250 yards below and east of a rock step now called the First Step. This was somewhat lower than Odell's sighting, but it almost certainly belonged to one of the two climbers. Then, in 1975, Chinese climber Wang Hongbao reported that he had spotted the body of a climber he referred to as "an old English dead." His grim discovery took place near Camp VI at about 27,000 feet. Unfortunately, he could not be questioned further because he died a day later on the mountain.

In 1986, American Tom Holzel engaged British Everest historian Audrey Salkeld for the first expedition specifically mounted to search for the pair. Their effort was unsuccessful. In 1999, a young German, Jochen Hemmleb, teamed up with Salkeld and an American team led by Eric Simonson to mount a second effort. Their discovery would shock the mountaineering world.

# "WHAT WE GET FROM THIS ADVENTURE IS JUST SHEER JOY. AND JOY IS, AFTER ALL, THE END OF LIFE."

–GEORGE MALLORY

THE LAST PICTURE of George Mallory and Andrew Irvine, taken on July 8, 1924, shows them at Camp VI at 26,800 feet (8,169 meters), leaving the North Col for their summit attempt. Noel Odell, who was in the second best climbing shape on the team after Mallory, took the shot. He would spend 11 days above the North Col searching for his friends—or for an explanation of what happened.

# EARLY AUDACITY

Nine years after the 1924 attempt, another team headed off. Among them was Eric Shipton. Born in 1907 in what was then Ceylon—today's Sri Lanka—Shipton was brought to London at the age of eight in order to "settle down." Yet school proved to be a disaster, for Shipton showed the classic signs of dyslexia. He eventually moved to East Africa to farm. Already a keen climber, he attracted the attention of another Briton farming in Africa—H. W. Tilman. Although he was older by ten years than Shipton, the two soon became climbing companions. The gregarious Shipton and reclusive Tilman seemed unlikely partners, but they are probably the most legendary pair in the history of mountaineering.

Shipton's exploits earned him an invitation to climb in the Garhwal Himalaya, which in turn brought him to Everest. Although three climbers of the 1933 team equaled the height record set by Norton in 1924, they did not reach the top. Shipton was sure that a smaller, lightly equipped team, moving quickly, would have a much better chance.

The next year Maurice Wilson, a British aviator with no mountaineering experience, became convinced that he could climb the mountain. Armed only with his inflated self-confidence, he made his way through Sikkim and Tibet, traveling in disguise because he did not have a permit. An extreme ascetic, he planned to live on rice water, ascend the Rongbuk Glacier, and climb the peak in three or four days. Luckily, he came across the 1933 team's food dump at Camp III, which he seems to have sampled. But what he ate wasn't enough to get him to the summit, and he died not far from the dump. Even Eric Shipton, the master of minimalism, had to concede that Wilson had taken this approach too far. Still, he admired his conviction. "It was not mountaineering, yet it was magnificent," he later wrote.

By contrast, the 1935 expedition was a cautious endeavor. The team was charged solely with finding a way onto the Western Cwm from the north side. Many felt disappointment that this group, led by none other than Shipton, wasn't going all the way. Shipton had a horror of large-scale affairs and delighted in keeping expeditions lean and mean, a "lightweight" approach that endeared him to future generations of climbers. His team managed to scale 26 peaks over 20,000 feet (6,000 meters), which he described as "a veritable orgy of mountain climbing." On the team was a 19-year-old Tibetan Sherpa named Tenzing Norgay, hired by Shipton in Darjiling. The 11th of 13 children, he was already smitten with Everest, and Shipton gave him his first chance.

The 1935 expedition collected a great deal of information that would contribute to future success on the mountain. It explored much of the surrounding area, gained a better understanding of the time window in which the mountain could be safely climbed, and proved the value of a small, nimble team. This expedition showcased Shipton as a model explorer.

A large expedition was defeated by storms in 1936, and in 1938 Tilman took charge. Like Shipton, he favored the lightweight approach. Tilman was not a tolerant man, particularly of teammate Noel Odell, who was attempting to conduct scientific experiments with a humidity machine. "The results were so unexpectedly various that one concluded the thing was only guessing," Tilman concluded. He further criticized Odell for his lengthy field notes, which he characterized as a full-length tome. Even Shipton got on his nerves, carrying with him the "longest novel that had been published in recent years."

Tilman was famously fussy about food supplies. He scoffed at jam, pointing out that the only valuable component was the sugar. Why not just take sugar, eliminating all

BRITISH MOUNTAINEER AND explorer Harold William Tilman keeps a pipe in mouth for a portrait. He would lead the 1938 British expedition on Everest. An astute judge of the mountain, he wrote that "a small party run on modest lines had proved itself as likely to reach the top as a large expensive one."

the scene during the Mount Everest Reconnaissance Expedition of 1951, led by Eric Shipton. During the expedition, the team noted two possible routes to climb neighboring Cho Oyu. The next year, Hillary would climb Cho Oyu to an altitude of 22,470 feet (6,850 meters) before descending when a good route couldn't be located.

that extra weight? Tinned food, commonly used at the time, was another waste of time in his opinion, since most of the weight was in the tin! He preferred real food: bacon, ham, cheese, butter, and eggs. In 1938 he managed to serve up a "full English breakfast"—bacon and eggs—all the way to the North Col. Despite the good food, the expedition was another failed attempt, one that Shipton referred to as a "vile waste of time." Then World War II intervened and priorities shifted.

## NEPAL OPENS ITS DOORS

In 1949, the Kingdom of Nepal opened its doors to foreigners. The following year, American climber Charles Houston and his father, Oscar, gained permission to travel into the Solukhumbu region to survey the south side of Mount Everest. They invited Tilman, and despite his misogynous attitude, he accepted, even though a woman was coming along. "Hitherto I had not regarded a woman as an indispensable part of the equipage of a Himalaya journey but one lives and learns. Anyhow, with a doctor to heal us, a woman to feed us, and a priest to pray for us, I felt we could face the future with some confidence." Their assessment of the south approach to Everest was gloomy. Houston thought the icefall might be "forced" but admitted it didn't look promising. Tilman was more suc-cinct: "Impossible. No route."

The Everest reconnaissance was Tilman's last foray into the world's highest peaks. A practical man, he assessed his years in the Himalaya, his current physical condition, and his level of motivation and determined that it was time to move on. Long and dangerous voyages to Patagonia and Greenland brought him many more years of adventure, and, perhaps fittingly, on November 1, 1977, he sailed with a crew in the direction of the Falkland Islands. He was never heard from again.

In 1951, Shipton was invited to lead another reconnaissance trip to the south side of Mount Everest. Accompanying a stellar group of Britons were several Kiwis, including Edmund Hillary. Almost immediately they were confronted with the dangerously unstable tangled mass of snow and ice known as the Khumbu Icefall. Shipton felt the way was too dangerous, but Hillary disagreed, saying, "I knew the only way to attempt this mountain was to modify the old standards of safety and justifiable risk and to meet the dangers as they came; to drive through regardless. Care and caution would never make a route through the Icefall." They managed to cross the icefall to the Western Cwm, but a huge crevasse spanning the entire valley turned them back.

## "YOU MUST FREE YOURSELF OF FEAR."

–DOUG SCOTT, ENGLISH MOUNTAINEER, FIRST TO CLIMB SOUTHWEST FACE OF EVEREST

## THEN & NOW THERMAL COATS

| 1963 | 2012 |
|------|------|

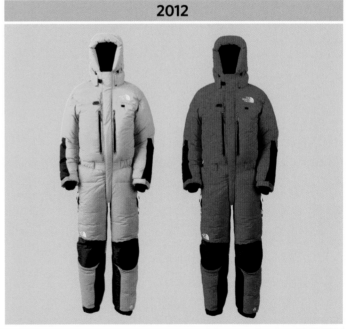

Parkas are a thing of the past for today's mountaineers, who wear lighter-weight, complexly designed full-body suits made for 8,000-meter peaks and extreme weather protection, such as The North Face Himalayan suit worn by National Geographic climbers in 2012.

Returning to Kathmandu, Shipton learned that the Swiss had secured a permit for the following year, meaning he would have to wait his turn. While the Swiss attempted Everest in 1952, Shipton had to be content with nearby Cho Oyu, which he and his team did not summit. The Swiss, meanwhile, very nearly reached the top of Everest. Raymond Lambert and Tenzing Norgay ascended the Southeast Ridge to within 800 feet of the summit. Shipton set off for an exploratory visit of the Barun and Arun Valleys, rather than returning home to take charge for the attempt the next year, and that decision would prove to be his undoing.

The Brits were determined more than ever to put someone on top in 1953. At headquarters, the committee fussed and fumed. They just *had* to get up this thing. What was wrong? After all these attempts, some key ingredient must be missing. Aha! It must be leadership. Perhaps a military man could organize the effort properly. They offered joint leadership to Shipton and Col. John Hunt, a man with limited high-altitude experience. Shipton resigned on the spot. It was clearly a slight—even an insult. But maybe the committee had a point. Charles Warren wrote that, despite his undying respect for Shipton, the decision was probably the right one. "I for one can understand why he was not eventually chosen to do so—the truth of the matter is that, by that time, his heart was not truly

## THEN & NOW CRAMPONS

| 1963 | 2012 |
| --- | --- |

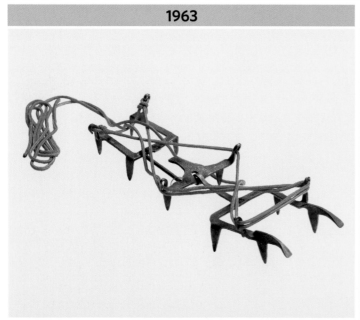
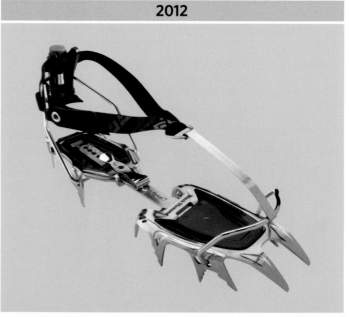

Crampons have transformed from a rather prehistoric look to a sci-fi elegance. Heavy-steel vintage models, used in the 1963 Everest expedition, rusted and had to be tied on with freezing, gloved hands. Today's precision-designed models feature serrated subpoints that grab and hold. The stainless-steel frames won't rust or collect snowballs, and the bootstraps clip on. They couldn't be more effective in tackling walls of ice.

After George Mallory was killed on Everest, his sisters secretly questioned whether their parents had any idea "what George was," whether they understood him and his wife Ruth at all. Now—when we have turned over every known word Mallory ever wrote, have retold his life story and reviewed his last climb endlessly, when we have probed even his sexuality—do we yet know what he was about? We remember him almost exclusively in mountaineering terms, but—compared to today's career mountaineers—the time he actually spent climbing was episodic. That's not to say he wasn't brilliant on rock, nor that he didn't derive intense spiritual sustenance from the hills and from the friendships he made among them. But you couldn't say Mallory "lived for mountains" when there was so much else in his life. It might repay us to take one further look at his story, putting aside the conventional acceptance that his premature death on that emblematic peak was the culmination of a personal quest . . .

## A CHARMING MAN

From the obituaries of Mallory, written by friends and others, you get a picture of a man with strong moral conscience, honesty, courage and a careless charm. Words most frequently employed by those who loved him were: sensitive, intelligent, chivalrous, impish and, as we would expect, fearless. Cottie Sanders rejoiced in his "appetite for beauty," saying no one ever had a greater genius for appreciation. Another word that recurs is "introspective." Geoffrey Young praised "his educational energy, his tenacious introspectiveness, that hunted to its source every interest but self-interest . . . " George was known affectionately as Sir Galahad; but was his Grail the highest mountain in the world, as it often seems, or was it a higher level of culture and understanding? We know that Ruth, who was a firm believer in an Afterlife, found it easier to reconcile her loss by believing that there was something better that he was ready for.

In many ways, Mallory was a slow developer. Though the gifts were there almost from the start, it took time for him to harness them. And that's what makes his death the more tragic: there is so much more we might have seen, had he lived. Frustratingly, it is still a hazy picture we have of what he planned through his work with the League of Nations Union or the Extra Mural Studies Board, and whether or not these two were interconnected. He passionately believed in the liberating effect of education, yet he may have gravitated towards politics—who knows? With his obvious charisma, even before the television era, we can imagine him reaching widening audiences via lecture halls and the radio. He could have been a candidate for popular programmes like *The Brains Trust*. It is impossible to envisage what impact he may have made throughout the thirties and the forties when he would have been at his career peak.

## HISTORY REMEMBERS

He himself, were he to look back at his abbreviated life, might fancy he had failed. Even his desire to be a writer had borne limited fruits in his lifetime, though now we have read almost every word he so prolifically got down in letters to family and friends—and we know he had the gift of communication that can leap generations.

More mementoes continue to turn up . . . As I write this, another letter, written a few days before he left that last time, is up for sale on eBay. An Everest ice-axe emerged recently, with the provenance of assistance given by Mallory to the emergent Pinnacle Club (a dedicated women's climbing club founded in 1921). Rumour hints at other intimate correspondences. Optimistically, I think we can expect to learn more.

—AUDREY SALKELD  **British author Audrey Salkeld has written numerous books and television scripts on mountaineering and exploration.**

George Mallory as a student at Cambridge, early 1900s

in it. Having discovered the route to the top by way of the South Col, he had really played his part, as the great explorer he was. For him it was the discovery that counted, not the conquest."

## TENZING AND HILLARY

When the British team requested the services of Tenzing Norgay the following year, he was initially reluctant to go, so loyal was he to Lambert. But the Swiss climber urged him: "Take the chance. It doesn't matter who it is with." Charles Evans and Tom Bourdillon came close to succeeding in 1953, reaching the South Summit at 8,750 meters, but their oxygen tanks were unreliable. The prize was left to the oddly matched couple: Edmund Hillary at six foot three and Sherpa Tenzing Norgay at five foot eight. They reached the top of Everest for the first time on May 29, 1953.

What a team effort it was. The day before, George Lowe, Alf Gregory, and Ang Nyima carried more than 40 pounds each, cutting steps all the way to the last camp at around 8,500 meters in support of Hillary and Tenzing. After dumping their loads, they left the two for their last night of preparation. At 6:30 the next morning, Hillary and Tenzing crawled out of their tent, hoisted their 30 pounds of oxygen gear, connected their masks, and started up. Things went smoothly until they reached a formidable-looking obstacle—the Hillary Step, as it would later be called. Hillary led the way and Tenzing followed. From there they just had to continue up the ridge until there was no place left to climb.

When the British reporter assigned to the team, James Morris, realized that the coronation of Queen Elizabeth II would take place three days later, he scrambled to get the news back to London in time. What a gift for the young queen! Written in code, the telegram transmitted the amazing news: "Summit of Everest reached on 29 May by Hillary and Tenzing." As the crowds gathered in the streets of London during the night of June 1, in eager anticipation of the royal parades the next day, the newspaper headlines screamed, EVEREST CLIMBED.

Morris called it the "last innocent adventure," but the international wrangling that occurred afterward almost destroyed the special relationship between Hillary and Tenzing. The Nepali government tried to convince Tenzing to claim he was the first to reach the summit. They even tricked him into signing a document stating that fact, though it was untrue. (The pair maintained their "reaching the summit together" message up until 1993, when Hillary announced that he had stepped up first.)

Three countries then claimed Tenzing as their own. He recalled that for the first 38 years of his life, nobody cared what nationality he was: Tibetan, Indian,

# "THE PULL OF EVEREST WAS STRONGER FOR ME THAN ANY FORCE ON EARTH."

—TENZING NORGAY

EDMUND HILLARY (FORE-GROUND) and Tenzing Norgay climb high on the mountain at Camp IV (following pages). On the morning of May 29, 1953, Hillary discovered that his boots had frozen solid and spent two hours warming them up before making the final ascent.

IN 1953, SIR Edmund Hillary and other climbers from the British expedition make the first ascent of Mount Everest; in May, Hillary and Tenzing became the first known to have reached the summit, pioneering the South Col route (opposite).

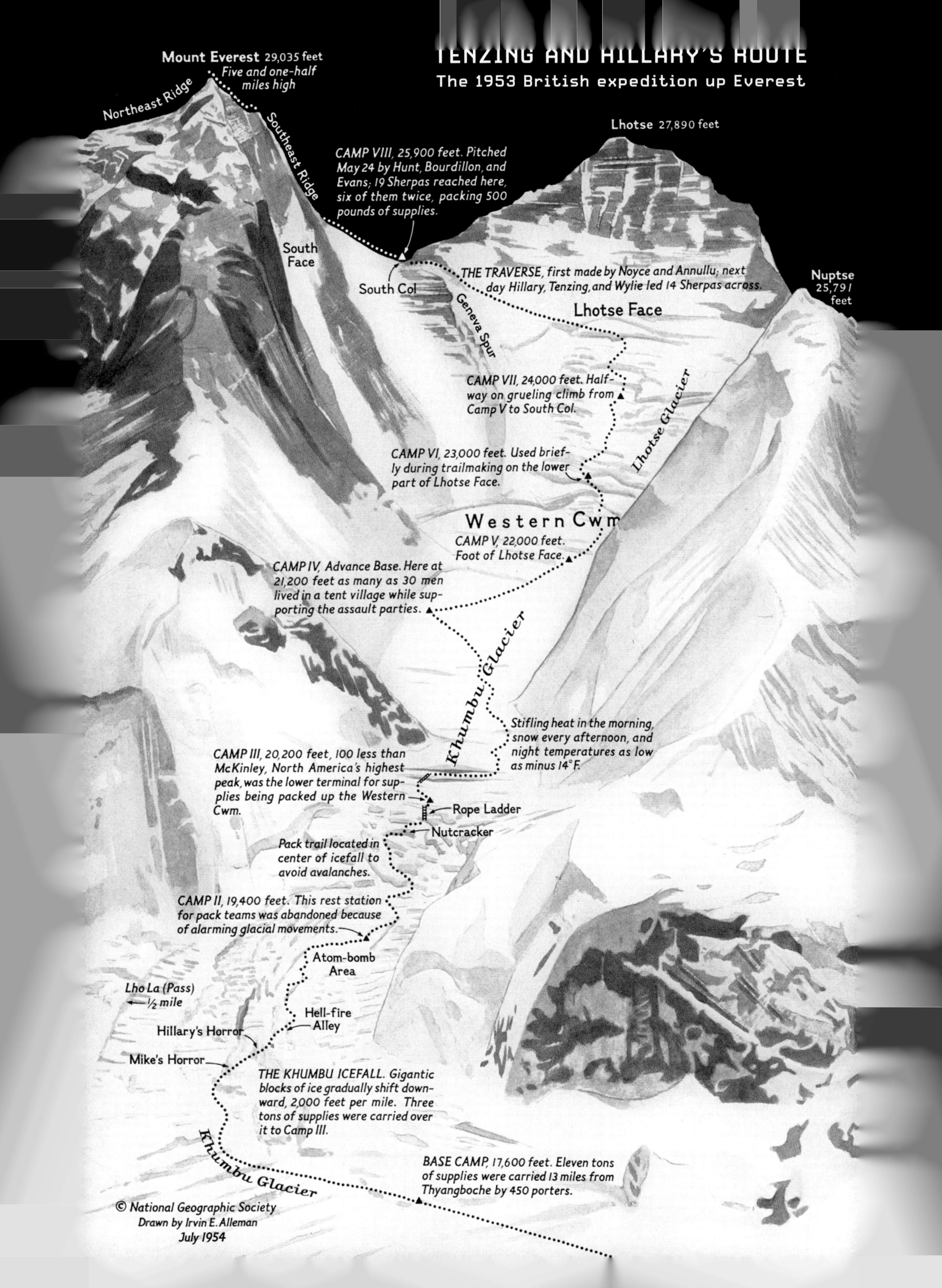

Mount Everest 29,035 feet
*Five and one-half miles high*

Northeast Ridge

Southeast Ridge

**TENZING AND HILLARY'S ROUTE**
The 1953 British expedition up Everest

Lhotse 27,890 feet

*CAMP VIII, 25,900 feet. Pitched May 24 by Hunt, Bourdillon, and Evans; 19 Sherpas reached here, six of them twice, packing 500 pounds of supplies.*

South Face

Nuptse 25,791 feet

South Col

*THE TRAVERSE, first made by Noyce and Annullu; next day Hillary, Tenzing, and Wylie led 14 Sherpas across.*

Geneva Spur

Lhotse Face

*CAMP VII, 24,000 feet. Halfway on grueling climb from Camp V to South Col.*

Lhotse Glacier

*CAMP VI, 23,000 feet. Used briefly during trailmaking on the lower part of Lhotse Face.*

Western Cwm

*CAMP V, 22,000 feet. Foot of Lhotse Face.*

*CAMP IV, Advance Base. Here at 21,200 feet as many as 30 men lived in a tent village while supporting the assault parties.*

Khumbu Glacier

*Stifling heat in the morning, snow every afternoon, and night temperatures as low as minus 14° F.*

*CAMP III, 20,200 feet, 100 less than McKinley, North America's highest peak, was the lower terminal for supplies being packed up the Western Cwm.*

Rope Ladder

Nutcracker

*Pack trail located in center of icefall to avoid avalanches.*

*CAMP II, 19,400 feet. This rest station for pack teams was abandoned because of alarming glacial movements.*

Atom-bomb Area

Lho La (Pass)
← ½ mile

Hell-fire Alley

Hillary's Horror

Mike's Horror

*THE KHUMBU ICEFALL. Gigantic blocks of ice gradually shift downward, 2,000 feet per mile. Three tons of supplies were carried over it to Camp III.*

Khumbu Glacier

*BASE CAMP, 17,600 feet. Eleven tons of supplies were carried 13 miles from Thyangboche by 450 porters.*

© National Geographic Society
Drawn by Irvin E. Alleman
July 1954

or Nepali. Suddenly everyone wanted him. "But now everything was pushing and pulling. I was no longer a man, but some sort of doll to be hung from a string. It must be I who reached the top first—a yard, a foot, an inch, ahead of Hillary. For some I must be Indian, for others Nepali. The truth did not matter."

Everest had an enormous impact on the lives of its first summiteers. Of all that reached the top, Hillary gave back the most to the country that made him famous. His creation of, and commitment to, the Himalayan Trust was ongoing, generous, and meaningful. Schools, monasteries, and hospitals were constructed throughout the Khumbu region.

Still, no accolades could approach those given to Tenzing Norgay. For millions, he became almost the manifestation of a god. His name became mythical, resonating as a talisman for future hope. When Hillary attended the unveiling of a statue honoring Tenzing's life in Darjiling in 1997, he said, "I have never regarded myself as much of a hero but Tenzing, I believe, undoubtedly was. From humble beginnings he had achieved the summit of the world."

**HILLARY AND NORGAY** were all smiles after their successful summit. The lives of both men would be forever changed because of their newfound fame. Hillary would comment, "I didn't have any concept of the reaction from the media and from the general public."

check the oxygen sets again. The flow rates seem all right. Turning to Tenzing, I say: "How do you feel?"

He just grins and waves his hand upward toward the ridge. I lead off once more, cutting steps. My ax work is still pretty rhythmical and relaxed; I've been chipping away for well over an hour, but, so far, I've avoided the kind of tension that can turn up a sore arm.

One flight of steps, then another, and another. We follow the ridge as it curves around to the right, wondering where the top can possibly be, or if it exists at all. I cut around the back of one crag, only to have a higher one stare me in the face. It seems endless.

Tiring, I try to save time on one stretch by skipping the step cutting and relying on my crampons. After a few yards I go back to my ax; the angle is still too steep, too dangerous. The zest we have known at the top of the rock step is draining away. Dully, grimly, I hack a route around still another knob.

## STANDING ON TOP OF THE WORLD

Suddenly I realize that the ridge ahead doesn't slope up, but down. I look quickly to my right. There, just above me, is a softly rounded, snow-covered little bump about as big as a haystack.

The summit.

One last question concerns me: is the top itself just a large, delicately poised cornice? If it is, someone else can have the honor of stepping on it.

I cut my way cautiously up the next few feet, probing ahead with my pick. The snow is solid, firmly packed. We stagger up the final stretch. We are there. Nothing above us, a world below.

I feel no great elation at first, just relief and a sense of wonder. Then I turn to Tenzing and shake his hand. Even through the snow glasses, the ice-encrusted mask, the knitted helmet, I can see that happy, flashing smile. He throws his arms around my shoulders, and we thump each other, and there is very little we can say or need to say.

My watch shows 11:30. Two hours and a half it has taken us from the South Peak; five hours from our tent. It seems a bit longer.

**Sir Edmund Hillary welcomed by Khunjung students, 1989**

## PHOTOS PROVE SUMMIT WAS REACHED

I turn off my oxygen and remove my mask. In the thin air of 29,000 feet my breathing becomes slightly more rapid, but not too uncomfortable. I fish out the camera I have kept warm inside my shirt; it will be necessary to take shots down every ridge if we're to prove conclusively that we've been up here . . .

Scooping a small hole in the snow, Tenzing buries a few offerings to the gods that many Buddhists believe inhabit these heights: a small blue pencil given him by his daughter, a bar of chocolate, some biscuits, a cluster of lollypops. I place near these gifts a little crucifix that John Hunt has received from a friend and passed over to me on the South Col.

It's time to go down now. I replace my oxygen mask, suck the air in gratefully, and move off without a backward glance. Reaction has set in; we both are tired.

We crampon along the steps I have cut, moving fast. We know the route; we know what's ahead and what isn't; the certainty gives us confidence and a lift to our stride . . . An ice-ax belay won't hold in the soft snow. If one of us begins to slide, both of us will enjoy a 10,000-foot jump without benefit of parachute.

—SIR EDMUND HILLARY
**A beekeeper from New Zealand, Hillary was knighted in 1953 and dedicated his life to philanthropy on behalf of the Himalaya.**

## NEW ROUTES

The next big breakthrough occurred in 1960, when the Chinese/Tibetans embarked on a massive 214-member expedition whose leader was none other than Mao Zedong. They chose the North Ridge, featuring the notoriously difficult rock steps where Mallory and Irvine were last sighted. The Chinese apparently solved the problem with a human ladder arrangement that, when they dispensed with boots and crampons, actually worked. But they were still a long ways from the summit. Not until 4:25 a.m. on May 25 did Wang Fuzhou and Qu Yinhua (Chinese) claim the summit with Gongbu (Tibetan). Their ascent wasn't acknowledged for years, a fact that Everest historian Audrey Salkeld chastised Western climbers for doubting. Skepticism was so entrenched that the Chinese felt compelled to do it again—in 1975. This time they brought a ladder for the steps, a feature that remains on the mountain to this day.

### AMEE '63

By the time the Americans arrived in 1963, Everest had been explored on two of its three main ridges: Only the West Ridge remained. But what the Americans needed most of all was to make the ascent. No American had done that, and it was high time. They were led by Norman Dyhrenfurth, a Swiss filmmaker with the looks of a film star and a personality to match, who had been the official photographer on the 1952 Swiss attempt. Sponsored by the National Geographic Society, the American Mount Everest Expedition (AMEE) team included Jim Whittaker, Barry Bishop, Lute Jerstad, Willi Unsoeld, and Thomas Hornbein. Logistics were handled by Nepal veteran Jimmy Roberts. This was an expedition destined for success. Victory was expected, not only from fellow Americans but also from the expedition's sponsors. Dyhrenfurth's initial plan was to do the grand slam: Everest, Lhotse, and Nuptse. But early on, Hornbein, an anesthesiologist from Seattle, and Unsoeld, deputy director of the Peace Corps in Nepal, began to imagine a grander plan—a new route on Everest.

Their dream presented daunting challenges for the team. Their discussions went on endlessly about where, and how, and when, and with what support. Democracy is a wonderful concept, but not necessarily when climbing a big mountain. Reaching the summit was imperative, so the South Col route, they decided, was the first priority. The West Ridge team would have to be content with what was left over. Once the summit was reached by the South Col route, the West Ridgers would get their moment.

CLIMBERS ON THE 1963 American Mount Everest Expedition (opposite top) leave advance base camp on their way to the summit, then work their way up the mountain (opposite bottom). On May I, Jim Whittaker, climbing with Sherpa Nawang Gombu, would become the first American to reach the summit. Four more team members summited on May 22, two along the West Ridge route and the others from the Southeastern Ridge. The expedition included more than 900 porters (below), who carried 27 tons of supplies to Base Camp.

Which they did. What made Hornbein and Unsoeld's such an iconic climb was that they reached a point of no return. They had only each other. Even after they had summited via what became known as the Hornbein Couloir, they then had to assist their teammates on the south side of the mountain in a frighteningly cold 28,000-foot (8,535-meter) open bivouac.

They weren't the first Americans to reach the summit. That honor was achieved by the South Col team, led by Jim Whittaker and Nawang Gombu. The two were followed by Lute Jerstad and Barry Bishop, who ended up sharing that desperate bivouac with the West Ridge boys near the top. They all survived, but the cost in digits was high: Unsoeld lost nine toes and Bishop lost all ten.

Everest changed the destiny of those 1963 climbers—certainly for Whittaker, whose social circle revolved around the Kennedy clan following his Everest triumph. Tragedy followed Unsoeld, who would lose not only his own life in the mountains, but also that of his daughter, Nanda Devi Unsoeld. Perhaps more reflective, Hornbein continued to search for the meaning of his Everest success: "Why was I here? I seemed to be hunting for answers to questions I couldn't even ask. What difference could Everest make even if I got to the top? What was up there to make me any wiser?" It would take him years to begin to discover the answers.

## THE LAST GREAT PROBLEM

Not until 1975 was another major route claimed on Everest, when Chris Bonington's Southwest Face expedition placed Dougal Haston, Doug Scott, Peter Boardman, and Pertemba Sherpa on the summit. But that landmark ascent was not accomplished on the first try. An earlier attempt to tackle this "last great problem" took place in 1971 under the leadership of Norman Dyhrenfurth, a two-time Everest veteran. It ended in defeat and interpersonal warfare. The second assault took place in 1972, this time led by Dr. Karl Herrligkoffer, who invited along British climbers Don Whillans, Hamish MacInnes, and Doug Scott. The three Brits managed to function under Herrligkoffer's famously authoritative style, but none of them was able to find a way to the top.

Chris Bonington led an unsuccessful attempt later that fall and returned in 1975 with a carefully chosen team. Dougal Haston, a Scotsman living in Switzerland, was a driven, ambitious alpinist. Martin Boysen was an old climbing

friend of Bonington's, and Doug Scott had already made an attempt. Nick Estcourt, Tut Braithwaite, and Mick Burke filled out the roster of this well-financed and superbly organized team. Peter Boardman described the effort best: "For a mountaineer, surely a Bonington Everest Expedition is one of the last great Imperial experiences that life can offer!" The key to their success occurred high up on the face. Previous teams had been defeated by the rock band when they had chosen a right-slanting ramp that dead-ended. This time they went left.

Dougal Haston and Doug Scott reached the summit at 6 p.m. on September 24, almost 16 hours after leaving their Camp VI. The photographs show a lovely—and ominous—sunset. Doug Scott enthused: "The view was so staggering, the disappearing sun so full of colour that the setting held us in awe." "Disappearing" is the pertinent adjective: It was much too late to return to their camp, so they ended up spending the night on the South Summit at 28,700 feet (8,750 meters), a new record for the highest bivouac.

Two days later Martin Boysen, Mick Burke, Peter Boardman, and Pertemba Sherpa headed up. Boardman and Pertemba reached the top. While Boysen turned back, teammate Burke continued on alone. He was last seen by Boardman and Pertemba resting just a few hundred yards from the summit. He seemed confident, but then the weather closed in, and Mick Burke disappeared forever.

**DOUGAL HASTON (OPPOSITE)** climbs through a snow-filled Hillary Step after his successful ascent. He had attempted the route two times before, only to be turned back before reaching the top. He is all smiles at the summit (above). In September of 1975, he and Doug Scott became the first Britons to reach the top of the world: They climbed using a new route, accessing the peak from the Southwest Face.

**JAPANESE CLIMBER JUNKO**
Tabei can't hide her joy after becoming the first woman to reach the summit, on May 16, 1975. She climbed with her Sherpa guide, Ang Tshering. Tabei was surprised by the difficulty of the final knife-edge ridge between the South Summit and the top. "I had never felt that tense in my entire life," she said.

Over on the Southeast Ridge, a diminutive piano teacher, only five feet tall and weighing well under 100 pounds, Japanese climber Junko Tabei became the first woman to climb the mountain. Just 11 days later, Phantog, a Tibetan mother of three, reached the summit with a Chinese expedition. Two women in two weeks! Another woman would not grace the summit of Everest for another five years.

## WITHOUT OXYGEN

Rumor has it that when Italian alpinist Reinhold Messner and an Austrian, Peter Habeler, were flying home in 1975 after climbing Pakistan's Gasherbrum I (26,469 feet/8,068 meters) without supplemental oxygen, they toasted their success with a gin and tonic and declared, "To Everest—without oxygen!" Three years later, they trained hard in order to climb quickly and efficiently on Everest, spending as little time as possible above 8,000 meters. Although Habeler expressed misgivings about dispensing with the $O_2$ while on the actual climb, Messner remained adamant. To save their breath, they scratched messages to each other in the snow rather than talk. At one point, Habeler's etched arrow

As we descended, the falling snow gave way to a fine drizzle. There was nothing to see; just one foot, then another. But slowly a change came, something that no matter how many times experienced, is always new, like life. It *was* life. From ice and snow and rock, we descended to a world of living things, of green—grass and trees and bushes. There was no taking it for granted. Spring had come, and even the grey drizzle imparted a wet sheen to all that grew. At Pheriche flowers bloomed in the meadows.

Lying in bed, Willi [Unsoeld] and I listened to a sound that wasn't identifiable, so foreign was it to the place—the chopping whirr as a helicopter circled, searching for a place to land. In a flurry of activity Willi and Barrel [Barry Bishop] were loaded aboard. The helicopter rose from the hilltop above the village and dipped into the distance. The chop-chop-chop of the blades faded, until finally the craft itself was lost in the massive backdrop. The departure was too unreal, too much a part of another world, to be really comprehended. Less than five days after they had stood on the summit of Everest, Barrel and Willi were back in Kathmandu. For them the Expedition was ended. Now all that remained was weeks in bed, sitting, rocking in pain, waiting for toes to mummify to the time for amputation . . .

### MISSION ACCOMPLISHED

We were finished. Everest was climbed; nothing to push for now. Existence knew only the instant, counting steps, falling asleep each time we stopped to rest beside the trail. Lester, Emerson, and I talked about motivation; for me it was all gone. It was a time of relaxation, a time when senses were tuned to perceive, but nothing was left to give . . .

We'd climbed Everest. What good was it to Jake [Breitenbach, who had died earlier on the climb]? To Willi, to Barrel? To Norman [Dyhrenfurth, AMEE team leader], with Everest all done now? And to the rest of us? What

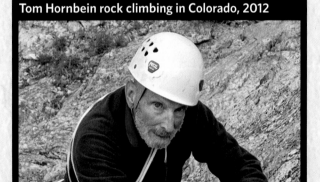

Tom Hornbein rock climbing in Colorado, 2012

waits? What price less tangible than toes? There must be something more to it than toiling over the top of another, albeit expensive, mountain. Perhaps there was something of the nobility-that-is-man in it somewhere, but it was hard to be sure.

Yes, it satisfied in a way. Not just climbing the mountain, but the entire effort—the creating something, the few of us moulding it from the beginning. With a lot of luck we'd succeeded. But what had we proved?

### THE SIMPLE LIFE

Existence on a mountain is simple. Seldom in life does it come any simpler: survival, plus the striving toward a summit. The goal is solidly, three-dimensionally there— you can see it, touch it, stand upon it—the way to reach it well defined, the energy of all directed toward its achievement. It is this simplicity that strips the veneer off civilization and makes that which is meaningful easier to come by—the pleasure of deep companionship, moments of uninhibited humour, the tasting of hardship, sorrow, beauty, joy. But it is this very simplicity that may prevent finding answers to the questions I had asked as we approached the mountain . . .

It had been a wonderful dream, but now all that lingered was the memory. The dream was ended.

Everest must join the realities of my existence, commonplace and otherwise. The goal, unattainable, had been attained. Or had it? The questions, many of them, remained. And the answers? It is strange how when a dream is fulfilled there is little left but doubt.

—THOMAS HORNBEIN

**Tom Hornbein, a member of the 1963 American team, scaled the difficult West Ridge with Willi Unsoeld. This passage comes from his book** *Everest: The West Ridge*, **published in 1965 and reissued in a 50th-anniversary edition in 2013 by The Mountaineers Books.**

# THEN & NOW CLIMBING ROPE

| 1963 | 2012 |
|---|---|

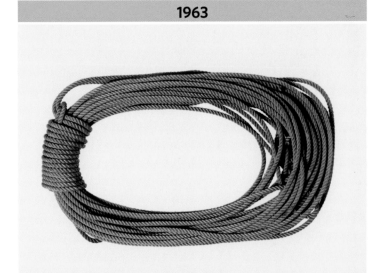

Ropes have vastly improved since the heavy hemp product used in the 1963 Everest climb. Ropes in 1924 were made of plant fibers like jute or hemp. Today climbing ropes are made of dynamic (stretchy) nylon known for its durability, light weight, and smooth handling.

pointed down. Messner's pointed up. Messner recalled the Herculean effort later in his *Expedition to the Ultimate:* "I am nothing more than a single, narrow, gasping lung, floating over the mists and the summits." Habeler was terrifyingly near the end of his limit when they reached the top: "It was a very personal, lonely victory in a struggle which each of us fought alone . . . In spite of my euphoria, I was physically completely finished."

Many believed their brains would be fried by the lack of oxygen, but they appeared quite lucid upon their descent. The Sherpas at Base Camp were astonished that they had done it. In fact, a strange situation arose in Kathmandu following the climb when a group of Sherpas called a press conference to denounce Messner as a liar. They didn't believe his claim to have climbed without oxygen, asserting that he had hidden tiny bottles of oxygen under his down jacket. If they couldn't do it, nobody could.

That same year, Wanda Rutkiewicz, Poland's first lady of climbing—and role model for the entire community of female Himalayan climbers—became the first European woman and the first Polish mountaineer to climb Everest. This beautiful, charismatic, and eagerly ambitious alpinist hated almost every moment of the climb due to some nasty interactions with her fellow climbers, something she attributed to male chauvinism. But her triumphant summit moment made up for the hardship, and her historic ascent coincided with Pope John Paul II's ordination day, Poland's greatest moment in the history of the Catholic Church.

A year later another Eastern bloc victory occurred: A strong Yugoslavian team led by Tone Škarja, the Chris Bonington of Yugoslavia, made a breakthrough on Everest when it climbed the entire West Ridge Direct. Never varying from the actual ridge, the team climbed overhanging rock above 8,000 meters—the hardest technical climbing on Everest to this point. Andrej Štremfelj and Nejc Zaplotnik finished the historic route and descended by the Hornbein Couloir. Their climb signaled the emergence of Yugoslavia (and later Slovenia) as a Himalayan powerhouse.

## EVEREST IN WINTER

From time to time, a visionary turns his attention to Everest. In the winter of 1979–80, that visionary was Andrzej Zawada, Poland's most charismatic expedition leader. Zawada had already established himself as a high-altitude winter specialist in the Hindu Kush as well as nearly ascending Lhotse in winter. Yet no 8,000-meter peak had yet been climbed in winter, and Zawada set his sights on Everest. With welding goggles to protect their eyes and homemade equipment to protect against the crippling cold and the screaming winds, his team had still not placed a climber on the summit when its permit was scheduled to run out in mid-February. Then Zawada managed to negotiate two more days, and two of the youngest, Krzysztof Wielicki and Leszek Cichy, reached the top on February 17. All of Poland rejoiced.

One of Poland's finest climbers, Jerzy Kukuczka, had missed the expedition because he had stayed home for the birth of his first son. Luckily, the insatiable Zawada had negotiated not one but two permits for Everest, and the second took place in the spring of 1980. Not content to repeat an earlier route, he plotted a

ANDRZEJ HEINRICH, A member of the 1980 Polish expedition, sports a frozen beard. On February 17 his teammates Leszek Cichy and Krzysztof Wielicki would become the first climbers to summit an 8,000-meter peak in the winter.

## VOICES — THE REASON FOR DANGER

The most important thing to know is that mountaineering involves risk. If we go to the mountains and forget that there is risk, we make mistakes. Mountains are dangerous! But they are only dangerous if people are there. A mountain is a mountain. It only exists. It is a piece of rock and ice, a beautiful piece maybe, but it only becomes dangerous and beautiful if we are there . . .

Danger is a filter that stops people from going where, perhaps, they should not go.

Danger has to do with managing fear, gaining experience, and learning hard-won lessons. Without making our own decisions, without accepting personal responsibility for our own actions, we cannot learn to achieve big things.

—REINHOLD MESSNER  The first person, with Peter Habeler, to summit Everest without supplementary oxygen, Reinhold Messner was also the first to climb all of the planet's 14 8,000-meter peaks.

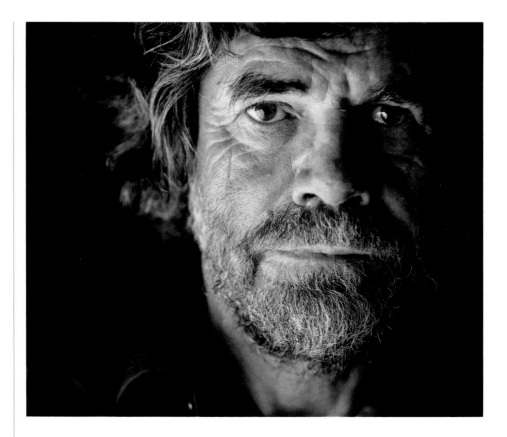

**ITALIAN REINHOLD MESSNER** (right), one of the world's greatest mountaineers, solo climbed to Everest's summit in August 1980 without using supplemental oxygen or a radio. He collapsed in his tent (below) after the feat, utterly spent. Messner is the first person to climb all 14 of the world's 8,000-meter peaks.

**CHOOSING THE LESS** traveled North Face route (opposite), Reinhold Messner made history in 1980 when he climbed Mount Everest alone without supplementary oxygen.

new line on the South Face between the South Pillar and the Southeast Ridge. High on the mountain, above 8,000 meters, a formidable rock band tested the very best: Jerzy Kukuczka and Andrzej Czok. Kukuczka recalled: "To climb this at that altitude took so much out of me that at one stage the effort made me simply wet my pants. At times my vision blurred."

## ACHIEVING THE IMPOSSIBLE

Reinhold Messner chose a spell of calm weather following the monsoon of 1980 to launch himself onto the North Face—alone. But the weather wasn't the only reason he opted for this particular time. In a casual conversation with Himalayan chronicler Elizabeth Hawley back in Kathmandu, he had learned that Japanese climber Naomi Uemura had plans for a solo climb of Everest. Messner advanced his schedule to ensure he would be there first.

He spent a month on the Tibetan Plateau, training his body to be content with little oxygen. But within several hours of setting off from his advance base camp at 6,500 meters, he almost lost his life when he fell into a crevasse. After extricating himself, he continued up the North Ridge and the next day traversed the massive, avalanche-prone North Face to gain the Great Couloir, where he bivouacked one more time. At 1 p.m. the following day he collapsed in the snow beside the Chinese tripod on the summit. He made the entire ascent without bottled oxygen, and this time alone. Dreadfully, frighteningly, alone. On his way down to his Canadian girlfriend, Nena Holguin, who was waiting at advance base, Messner realized that he had reached an apex in his career that none of the remaining 8,000-meter peaks that he would climb would ever match.

## THE ESSENTIAL SHERPAS

By now everyone realized that Sherpas were integral to most people's success on Everest. They were incredibly strong at altitude and seemed capable of withstanding intense cold, large loads, and long days. Jim Whittaker, the first American to the top, joked about their capacity at altitude: "You don't notice them until they take a deep breath. Then their lungs fill up, and they block the view."

Sherpas are inextricably interwoven into Everest history, and the record of their accomplishments on the mountain is legendary, although not often celebrated. Everyone knows of Tenzing Norgay, but hundreds of others could justifiably be called Tigers of the Snows. Dawa Tenzing was still carrying loads to the

# "IF PEOPLE WANT TO CALL ME 'THAT CRAZY MOUNTAIN WOMAN,' THAT'S O.K."

–JUNKO TABEI, 1996

Mount Everest
8,850 m
29,035 ft

Northeast Ridge

West Ridge

North Ridge

Reinhold Messner's first solo ascent

Messner's second bivouac

Changtse
7,583 m
24,879 ft

Messner's first bivouac

On May 19, 2012, Canadian Shriya Shah-Klorfine added her body to the 200 or more corpses that litter the flanks and ice fields of Mount Everest. Three others died that day, old and young, men and women, a German, a South Korean, and a Chinese. Urged by her Sherpa guides to turn back, Shah-Klorfine insisted on reaching the summit, which she did, though it meant her death on the descent. Still, undeterred by the tragedy, that season another 200 climbers from a dozen nations strapped on their boots and eagerly walked to heights where the air is so thin that humans cannot long survive. Each season, the pattern remains the same. Most will return. Some may not. For every 30 climbers who have reached the summit of Everest, 1 has perished in the quest.

## THE RISKS ON EVEREST

What are the ingredients of this perfect storm that brings death to the mountain every spring? First, there is the pressure of time. The climbing season on Everest is short, a mere two months squeezed between the end of winter in late March and the onset of the monsoon in early June, with the optimal window for summit attempts being but

Long lines encumber the ascent of the Lhotse Face during the 2012 climbing season. Simone Moro took this picture shortly after he decided to abort his climb due to the dangerous lines.

three weeks in May. Second, there is the structure and geography of the mountain. The highly technical routes on Everest—the North Face, West Ridge, the legendary Kangshung Face—defy all but the most accomplished of Himalayan climbers. For commercial expeditions, and for the vast majority of climbers, there are only two viable options. Those who climb from the south follow the tracks of Edmund Hillary and Tenzing Norgay, who in 1953 trekked through Nepal, made their way through the notorious Khumbu Icefall, and climbed the ice face of the Western Cwm to reach the South Col before enduring a brutal final ascent that famously took them to the summit of the mountain. From Tibet, climbers follow the footsteps of George Mallory and Oliver Wheeler, the first Canadian on Everest. It was Wheeler who in 1921 discovered the doorway to the mountain from the north, up the East Rongbuk Glacier to the North Col, and from its crest a long slog up the shoulder of the North Ridge to the Northeast Ridge and ultimately the base of the summit pyramid.

Neither conventional approach is easy, and each has unique perils. In 1922 seven Sherpas, swept to their deaths in an avalanche, became the first of many to perish on the face of the North Col. On the Nepal side the Khumbu Icefall is a maze of giant seracs that collapse randomly in the heat, crushing climbers with disturbing regularity. These hazards aside, the challenges of both popular routes are for the most part less technical than physical. Few of those strung out along the length of the Western Cwm, plodding their way to the South Col, would consider attempting K2, Nanda Devi, Kangchenjunga, Chomolhari—all far more difficult Himalayan peaks. On Everest, raw endurance counts for more than skill with a belay. Long before any client sets out for the summit, the guides in support establish a continuous ribbon of fixed lines stretching the entire length of the route; the client need only clip in and trudge up a

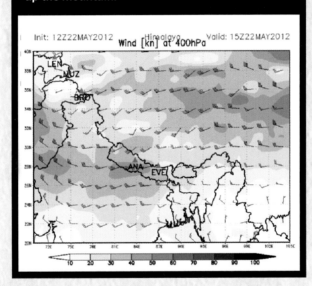

Meteotest's weather chart, based on data by the NOAA—the global forecast system that helps climbers predict the optimum weather window. Team leaders study this kind of chart when determining whether or not to continue higher up the mountain.

well-trodden track. Easier said than done, of course.

The danger is the very wrath of the mountain, the extremes of elevation and weather. Frigid nights give way to blistering days, and with the scorching sun, there is a constant danger of dehydration. As Charles Howard-Bury, leader of the 1921 expedition, famously quipped, on Everest "your feet can be suffering from frost-bite while you are getting sunstroke at the same time." Then there is the matter of oxygen. At 8,850 meters, atmospheric pressure at the summit of Everest is a third that of sea level. For the early British explorers, this distinguished the challenge of Everest from the quest for the Poles. It was one thing to face conditions of extreme cold and bitter exposure. It was quite another to do so while moving not laterally across landscape but vertically to heights where the very air itself could not sustain life. Climbers call it the death zone. Every minute one spends there weakens the body and increases the likelihood of injury or death.

## OVERCROWDING

Hence the tragic events of that weekend in May that led to the death of four climbers. Each of the approaches taken by commercial parties has one serious impediment. The Northeast Ridge is blocked by the Second Step, a formidable pitch of vertical rock that must be scaled even as the climber peers down to a knife-edge of ice and rock—exposed on one side to the Kangshung Face, a drop of 10,000 feet, and on the other to the North Face, a mere 9,000 feet. On the southern route is the Hillary Step, equally daunting with similar exposures. Today, fixed ropes on the Hillary Step and a ladder on the most difficult section of the Second Step make things easier, but neither obstacle can be turned, and on each there is room for only one person at a time. With as many as 300 climbers making their summit attempts in a single day, both routes take on the feel of traffic jams. The

Sherpas install ladders on Everest slopes.

Hillary Step and the Second Step become bottlenecks, literal choke points where climbers can end up spending hours in the death zone, exposed at high altitude, waiting their turn. According to reports from the mountain, on the day that Shriya Shah-Klorfine died, delays stretched to over three hours even as the weather turned, leaving exhausted climbers exposed to severe cold and winds gusting to 80 miles an hour. It is remarkable that only four were lost.

I often wonder what the early Brits would have thought of today's rather sordid commercial scene. In 1921, Mallory and Wheeler had to walk 500 kilometers off the map across the Tibetan Plateau just to reach the base of a mountain that no European had embraced at close quarters. When Wheeler first crested the North Col, he encountered a wind unlike anything he had known. Scarcely able to stand, fearful of suffocating in the swirling eddies of snow, he focused on his breathing, drew his hands around his face, and with a discipline long ago honed in the terror of shellfire on the Western Front slowed down the world until a new rhythm could be found and air inhaled during the lulls between the blasts of the gale.

They soon came to know what the conquest of Everest would demand. To the horror of the old guard at the Royal Geographical Society, codgers who for the most part had never climbed higher than their desks, George Finch in 1922 argued that the "margin of safety must be narrowed down, if necessary to the vanishing point." A climber on Everest must drive himself beyond exhaustion, "even to destruction if need be."

"We must remember," George Mallory wrote on the eve of his return to Everest in 1922, "that the highest of mountains is capable of severity, a severity so awful and so fatal that the wiser sort of men do well to think and tremble even on the threshold of their high endeavor."

## WITNESS TO THE OBSESSION

My thoughts turn also to those who have born witness to our obsession with the mountain. I once met a Buddhist living in a nunnery at the base of Everest who had spent 45 years in isolated retreat, dedicating her entire life to the recitation of a single prayer. To those prepared to sacrifice

Andy Bardon checks his altimeter on the 2012 Everest expedition.

everything in a quest to reach the summit of the world, such spiritual devotion may seem like a waste of a human life. Most Tibetans find it equally incomprehensible that one would choose to walk to heights where the air is so thin that consciousness is obliterated. To enter a death zone deliberately, to risk losing the opportunity of personal transformation and escape from the realm of samsara merely to climb a mountain, is for them folly, the actual waste of a precious incarnation. As the abbot of Rongbuk Monastery wrote of the British expedition of 1922: "They camped at the bottom of the mountain, then, I heard they camped for seven times for each level they reach, with great effort they use magical skills with iron nails, iron chains and iron claws, with great agony, hands and feet frozen . . . [Some] left early to have limbs cut off, the others stubbornly continue to climb . . . I felt great compassion for them to suffer so much for such meaningless work."

—WADE DAVIS **Wade Davis, an explorer-in-residence at the National Geographic Society, is an author, photographer, and cultural anthropologist. His most recent book,** *Into the Silence,* **is a cultural history of the British Everest efforts of the 1920s.**

# "I HAVE NEVER MET PEOPLE MORE ADMIRABLE IN THEIR WILLINGNESS TO DO WHAT IS RIGHT THAN THE SHERPAS."

—PRESIDENT JIMMY CARTER, 1985

**AMERICAN ED WEBSTER** (opposite) crosses through the free space over a huge crevasse using a Tyrolean traverse. In 1988 he and three teammates would attempt the summit via the Kangshung—or East—Face, without Sherpa support, supplemental oxygen, or fixed ropes.

South Col at age 56 for the Americans in 1963. Nawang Gombu, who reached the top with Jim Whittaker in 1963, later replaced Tenzing Norgay as the director of field training at the Himalayan Mountaineering Institute in Darjiling. Pertemba Sherpa climbed the Southwest Face with Pete Boardman on Chris Bonington's landmark 1975 expedition. Apa Sherpa, nicknamed "Super Sherpa," has reached the top 21 times. Babu Chiri Sherpa, who climbed the mountain by several routes and summited ten times—twice within a 14-day period—once spent the night on top just because he wanted to, and climbed it once in 16 hours and 56 minutes. The list goes on and on.

But not every team climbed with Sherpas. The Poles couldn't afford them. Nor did a small British team of eight led by Alan Rouse that tried Everest in the winter of 1980–81, this time by the West Ridge, with no supplemental oxygen. It was an incredibly ambitious undertaking, and they were turned back by the cruel winter. Though nobody died on this expedition, three of the eight would make the ultimate sacrifice in the Himalaya in the coming years: Joe Tasker on Everest in 1982, Pete Thexton on Broad Peak in 1983, and Al Rouse on K2 in 1986.

Bonington returned in 1982 with an all-star team of six, intending to climb the Northeast Ridge. Pete Boardman and Joe Tasker, well known in Britain not just for their climbing prowess but also their writing and filmmaking talents, were among them. The pair headed up the Northeast Ridge to an area called the Pinnacles at about 27,000 feet (8,200 meters). They were last sighted around 9 p.m. on the night of May 17, silhouetted just below the Second Pinnacle. The British casualty rate in the Himalaya was mounting. As historian Elizabeth Hawley observed, "A whole generation of British climbing has been decimated . . . Bonington is alive, but most of his friends are gone." Only in 1995 would the entire Northeast Ridge be climbed, by a Japanese team from Nihon University, using full siege tactics, including 13 climbers and 31 Sherpas.

At the same time that Bonington's team ascended the Northeast Ridge, a Soviet team of 17 forced a bold, new, and difficult line on the buttress to the east of the British Southwest Face route. Eleven climbers reached the summit, heralding the Russian style of big-wall, Himalayan climbing.

Finally, in 1983, the last of Everest's faces was climbed—perhaps the most intimidating and dangerous one—the East (Kangshung) Face. It was so hazardous that the American team chose not to use Sherpas, reasoning that it wasn't right to expose them to so much danger. Using supplemental oxygen, six climbers summited: It was the second new route on Everest pioneered by Americans. With this ascent, all three of Everest's great faces—North, Southwest, and East—had been climbed.

**STEPHEN VENABLES (ABOVE)** descends from the summit to the South Col after the first ascent of the Kangshung Face. Reaching the summit by himself, he became the first Briton to summit without using supplemental oxygen. On his descent, he spent a night in a bivouac at 28,000 feet (8,600 meters).

**DICK BASS MAKES** his way through the Khumbu Icefall (opposite). His 1985 ascent marked a turning point for Everest when he became the first guided client to reach the top. At that time, the 55-year-old Bass also became the oldest person to summit. It was his fourth attempt.

Two years later, another Everest first took place, and it presaged the future of climbing on the mountain. When Snowbird Ski Resort owner Dick Bass summited Mount Everest on April 30, 1985, he could fairly be classified as the first fully guided Everest client. It was his fourth try at the mountain, and he had David Breashears to thank for his success. "You got me up—and I know you'll get me down," he gasped to Breashears on the summit. Breashears did just that, for he fully understood the relationship. "He's my responsibility: I have to get him down."

With the explosion of guided climbs that followed, many have criticized the business of guiding clients on the highest mountain on Earth. Guiding is an honorable profession with a long and illustrious history, particularly in the Alps. But it's a different story above 8,000 meters. There are so many additional factors: the massive scale, the crippling altitude, the fierce storms and the huge costs. Perhaps most important, the guide has higher stakes as well. Each is granted only one summit per season, not like in the Alps. It's important to make each season count.

## CLIMBING NIGHT NAKED

In the summer of 1986, possibly the most impressive ascent of Everest was made by Swiss climbers Erhard Loretan and Jean Troillet. After acclimating for five weeks, they left their advance base camp together with French climber Pierre Béghin. They began at 11 p.m. on August 28 and struck out for the Japanese couloir on the north side. Climbing through the night, they stopped at 11 a.m. at 7,848 meters to relax and rehydrate in the warmth of the day. As the evening cooled, they started up the Hornbein Couloir. After four hours of climbing, they stopped to wait for dawn's light. Béghin turned around at this point, but the other two resumed their climb, reaching the summit at 2 p.m. After 90 minutes they started down, and since the snow conditions were superb, they chose to descend the entire face in a sitting glissade that took a mere four and a half hours, for a total of 39 hours round-trip. They took no tents, ropes, harnesses, or bottled oxygen. Polish climber Voytek Kurtyka (and frequent climbing partner of Loretan) called it "night naked" climbing. When the author asked Loretan, years later in 2010, what his most memorable climb was, he cocked his head, flashed a lopsided grin, and answered, "The Everest climb was very good . . . it was such a straight line."

The year 1988 was a conflicted one on Everest. Several amazing ascents were made, but in very different styles. A massive undertaking called the Japanese-Chinese-Nepalese Friendship expedition, 254 members strong and costing seven million dollars, sent climbers up the Southeast and Northeast Ridges simultaneously, broadcasting live from the summit.

In contrast, a small four-man international team was attempting a fierce new route on the East (Kangshung) Face via the Southeast Rib. Britain's Stephen Venables, Americans Ed Webster and Robert Anderson, and Canadian Paul Teare hardly knew each other, yet they hoped to make the ascent without fixed ropes, without supplemental oxygen, and without Sherpa support. When Charlie Houston first heard of the plan, he bellowed: "Four against the Kangshung Face? You're mad!" Judging by the features on their route, which they eventually called the "Neverest Buttress," Houston was probably right. Filled with horrific overhanging snow and ice mushrooms, vertical ice walls, and gaping bergschrunds (where a glacier splits off from its base), the route was dangerous, steep, and sustained.

They worked their way up through the desperate terrain all the way to the South Col. Then fatigue began to take

# "THESE MEN WERE GIANTS, LARGER THAN LIFE."

-BRENT BISHOP, SON OF BARRY BISHOP, MEMBER OF THE 1963 AMERICAN MOUNT EVEREST EXPEDITION

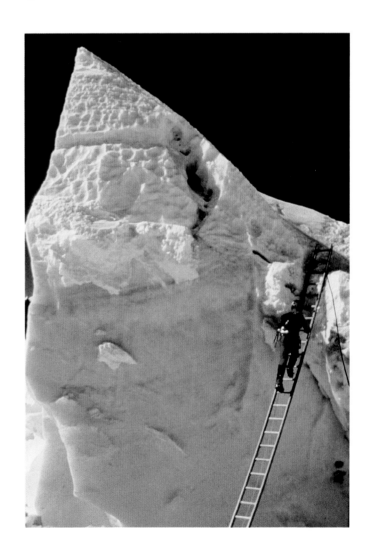

**Kit DesLauriers climbing the summit ridge of Mount Everest**

Shooting a beautiful image on Everest isn't hard. Being on Everest is hard. As a photographer shooting an expedition, the true challenge is to create a diverse montage of the experience peppered with emotional, storm-swept, descriptive, eye-catching, eye-keeping, storytelling images. I often have a hard time with those challenges at sea level. Bring those challenges up over 25,000 feet and it can be a bit overwhelming.

Expedition shooting poses an interesting dilemma. There is the constant dialectic about whether one should focus on the climbing or the creative. One's comfort level with the terrain and climbing often dictates one's ability to shoot and be creative. I have a love-hate relationship with this challenge. I have built a career out of it. I have dreaded it. I have found some of my greatest moments of satisfaction in it. And it has almost killed me.

## BE PREPARED

In general, I know that at any given moment there is always an amazing image to be created, but you can't shoot all the time. You need to pick your moments, anticipate them, work them.

When the light is happening and there is a flurry of action, you'd better be prepared. Your food for the day and camera gear need to be prepped; you need to be dressed before everyone and out the door so you can shoot everyone coming out. You need to have your bag packed so you can shoot everyone else packing but still not fall behind when they leave, because you need to get ahead for the next shot! Shooting people from below with a big blue sky above is rarely interesting. Shooting from above to show the massive exposure of a sweeping face below them and the entire Himalayan range spread across the horizon is the shot you want.

—**JIMMY CHIN** **An unparalleled expedition filmmaker and photographer, Jimmy Chin is also a climber, mountaineer, and explorer in his own right. He is a National Geographic emerging explorer and a member of The North Face athlete team.**

its toll and Venables pulled ahead, alone. As he climbed through the night on May 11, he first assumed the others would soon catch up to him. Eventually it became clear that he was on his own: Either break trail to the summit—all the way—or admit defeat. He made the summit but failed to reach the shelter of the tent on the South Col on his return. Instead he bivouacked, hallucinating wildly, with visions of Eric Shipton warming his feet and Tibetan yak herders beckoning him to their fire. At dawn he staggered down to the tent, where two of his three mates huddled.

But the story didn't end there. After a frightening night, crammed together at the South Col and Venables half frozen, the climbers had to decide which route to descend: back down the Kangshung Face, or the easier South Col route. Despite its ferocious difficulty, they chose the face because they knew the way. After a series of wild free-falling glissades and miraculous recoveries, all but one had lost their ice axes. The story of their three-day descent, so completely out of control, is not a model to be repeated. Nevertheless they survived, garnering acclaim. Reinhold Messner called it the "best ascent of Everest in terms and style of pure adventure." Chris Bonington was more specific: "Amongst the most remarkable examples of survival in the history of Himalayan mountaineering." Survival it was, but not without cost. Webster lost the tips of seven fingers and a thumb, and Venables lost three and a half toes.

That autumn, on the south side of the mountain, a small team of Slovakia's best climbers tried to make the first alpine-style ascent of the Southwest Face, initially climbed in 1975 by Chris Bonington's team. They used no fixed ropes, supplemental oxygen, or Sherpa support. The Slovaks succeeded on the face, but all four died on the descent. Because of the devastating loss, many began to question just what defined "success."

At the same time, Lydia Bradey, a flamboyant young New Zealand climber, was climbing with a New Zealand team that shared the Southwest Face permit with the tragic Slovak group. The New Zealand team, led by Rob Hall, did not reach the summit and returned to Kathmandu. All except Lydia. She climbed to the summit illegally via the Southeast Ridge without supplemental oxygen—making her the first woman to do so. The problem was that she had no watch to record the time, and her frozen camera had malfunctioned. She had no proof. Rob Hall dismissed her claim, stating that it was simply not possible. Yet he had an

IN AUGUST 1986, Jean Troillet and the late Erhard Loretan (pictured here) set out to climb Everest—fast. They summited without supplemental oxygen or Sherpa support in just 39 hours using the Hornbein Couloir.

ulterior motive: Her illegal ascent might jeopardize his ability to guide in Nepal in the future. The summit claim was listed as "disputed," thereby depriving Lydia the accolades for what should have been a monumental achievement. Only years later, after confirmation by a Spanish climber on the mountain, was Lydia Bradey finally credited with the first female ascent of Everest without supplemental oxygen. Lydia was one of the few "firsts" on Everest for whom the accomplishment did little to change her life. She was a mountain guide when she climbed the mountain, and a guide she remained.

# THE RECENT PAST

Almost every decade has offered up unique and creative advances on Everest: The 1920s and '30s were devoted to exploration; the first summit successes came in the '50s; new ridge routes began in the '60s; the '80s saw new routes and oxygen-free ascents. But what about the '90s? According to American climber Ed Webster, "In the early 1990s the years of wonder on Everest seemingly ended."

Part of the reason is the vast increase in the number of climbers. The Nepal Ministry of Tourism sells more and more permits for the Southeast Ridge each year—a veritable gold mine. China has followed suit for the North Ridge. These are granted to climbers chasing the Seven Summits. Climbers chasing the 14 8,000-ers. Climbers chasing Everest. Many, including Messner, feel that Everest is no longer a true mountaineering experience. British author Ed Douglas called it "Himalaya Horribilis." Thousands now throng the mountain via its most popular routes, and the vast majority of them are guided. Hundreds reach the summit each year feeling triumphant, for Everest has been their dream, their ambition, their obsession. Yet many others continue to die trying, some from falls, others from hypoxia, and most from sheer exhaustion.

## THE YEAR OF DISASTER

The year 1996 was when too many died. The highly publicized disaster that killed a total of 11 people has sadly defined Everest in the 1990s. Many have analyzed the sequence of events that led to the tragic night. Fingers have been pointed. But as with most mountain tragedies, many factors contributed. Certainly, there were too many

**DAVID BREASHEARS (FOREGROUND)** and members of his team brave near-whiteout conditions to retrieve the body of Chen Yu-Nan. The Taiwanese climber had fallen into a crevasse on Lhotse Face and succumbed to his injuries. He would be just one of 15 fatalities that year, 1996.

people crowding the route on summit day, creating gridlock and terrible delays. Communication problems created confusion about who was supposed to do what, and at what time. Some guided clients weren't as experienced as they should have been, for they would have relied less on fixed lines and hand-holding on the epic descent. The rivalry between the leaders of the two commercial expeditions—Rob Hall, a New Zealand guide and owner of Adventure Consultants, and American Scott Fischer, head of Mountain Madness—probably also played a role. Fischer was the new guy on the hill, and Hall, the well-established Everest guide, had succeeded in getting his clients to the summit every year, except for the previous year. The pressure must have been fierce. Hall disobeyed his own rules of survival when he and his clients stayed too long above 8,000 meters. The year before he had followed those rules and, although nobody reached the summit, everyone had returned from the climb. In 1996 four of his team, including Hall himself, perished on their descent.

Fischer was ill on the day his group summited, compromising his ability to look after his clients and bringing into question the strategy that he had

**ADVENTURE FILMMAKER LEO** Dickinson (below) has a bird's-eye view of Everest on the first balloon flight over the peak. The balloon cleared the summit by about 3,000 feet (900 meters) in a flight that lasted about an hour and a half. Dickinson recorded the flight in spectacular photographs.

**BRITISH CLIMBER ALISON** Hargreaves tackles a steep pitch during her successful 1995 attempt to become the first woman to summit Everest unaided and without supplemental oxygen (opposite). She was only the second person, after Reinhold Messner, to summit unaided.

developed with his Russian guide, Anatoli Boukreev. Boukreev was to guide from the front, and Fischer from behind. That strategy couldn't succeed with an ailing Fisher. The clients needed more support.

Other climbers on the mountain observed that all the guides were clad in relatively lightweight clothing, indicating perhaps overconfidence in their ability to get up and down the mountain quickly.

The costs associated with guided climbs of Everest probably influenced some clients to climb beyond the point of exhaustion. They couldn't afford to return.

The most significant factor was, of course, the storm. According to some, it was the storm of the century. Others rated it just another storm on Everest. No matter how fierce, it combined with all of the individual elements leading up to summit day—illness, broken rules, miscommunication, lack of skills, exhaustion—to become the perfect storm. Nine people lost their lives that night. Even more would have perished had Boukreev not gone out into the night, again and again, to rescue climbers huddled in the blizzard on the South Col. Rob Hall was not so fortunate. After he helped a climber reach the summit, his oxygen tank froze, and he died in the blizzard.

# VOICES BESIDE MALLORY

Sitting beside Mallory on May 1, 1999, I looked east toward the descent route he and Irvine would have taken had they summited that June 8, 1924. I imagined Mallory's awareness even in extremis: no radio to communicate with others, no chain of fixed ropes to guide him down the mountain, no teams of rested climbers ready to enact a rescue, no way of telling the world what really happened.

I can only guess what Mallory's and Irvine's last moments were like but what I do know is how their achievement has affected our climbing legacy. The boldness of their last climb formed a stepping-stone to the future. The debate over oxygen and its ultimate acceptance made it possible for their successors—including Hillary and Tenzing—to visit high places with a reasonable safety margin.

Sometimes late at night I wonder whether by discovering Mallory I've aided in the destruction of a mystery.

The possibility haunts me. Has my find somehow taken some of the enigmatic glory away from the 1924 expedition?

Others may think so, yet for me, the discovery only increases my admiration for these pioneer climbers, whose story—which will never be told in its entirety—has always lain wrapped in the secrets of Chomolungma, "mother goddess of the snows." I feel privileged to have participated in casting new light onto this mystery. Ultimately, Mallory and Irvine's greatest achievement was an inspirational one, for even in failure, their magnificent attempt showed us what the human spirit is capable of.

—CONRAD ANKER Conrad Anker is recognized the world around as a premier alpinist. With funding from the National Geographic Society and The North Face, he led the 2012 Legacy Climb up Everest to commemorate the first American team's summit 50 years before.

## PIONEERS FOREVER

Despite the tragedy that sadly dominates this decade, many others took on the challenge of Everest and succeeded, sometimes in very original ways. At the beginning of the decade, Australian Tim Macartney-Snape began his journey at the Bay of Bengal and continued, walking (and sometimes swimming) the entire distance to the mountain's base, and then to the summit. In 1990 Jim Whittaker led the Peace Climb, a joint venture between the Chinese, Americans, and Soviets that placed 20 climbers on the summit. Also that year, the first married couple climbed to the summit when Maria and Andrej Štremfelj, two veteran Slovenian climbers, barely beat a rival team, American Cathy Gibson and her Russian husband, Aleksei Krasnokutsky.

In 1991, British filmmaker Leo Dickinson ballooned over Everest, taking some of the most superb photographs ever seen of the mountain. On April 22 of the following year Pasang Lhamu became the first Nepali woman to climb Everest—on her fourth attempt at the mountain. But her story did not turn out happily. While she became a national hero, she died on her descent and took a Sherpa with her—Sonam Tshering, father of three children and an unborn

fourth. A somewhat similar fate greeted British climber Alison Hargreaves, who in 1995 climbed the mountain alone without supplemental oxygen. Sadly, she would lose her life in August of that year as she descended from the summit of K2.

The same year as the 1996 *Into Thin Air* disaster, veteran Everest climber and award-winning filmmaker David Breashears made the first IMAX film on Everest. That same year, a South African team, including Bruce Herrod, Cathy O'Dowd, and Ian Woodall, placed the South African flag on the summit for the first time. But Bruce Herrod lost his life in the process. Everyone on the mountain stopped to listen when President Nelson Mandela called their base camp to offer congratulations to the summit climbers. When he learned they were still high on the mountain, he asked that they call him back, and even began to give his telephone number, over national radio! Luckily, their base camp manager cut that particular transmission short, to the amusement of the entire mountain.

Also in 1996, a strong Russian team from Siberia led by Sergei Antipine climbed a bold new line straight up between the North and Northeast Ridges of the mountain, succeeding on their very first attempt. The slopes varied from 65 to 90 degrees of steepness.

In the same year Swedish climber Göran Kropp rode a specially designed bicycle, groaning with its 240 pounds of equipment, 7,000 miles from Stockholm to Kathmandu, then carried on to the summit of Everest. He survived the deadliest Everest season only to die rock climbing near his Seattle home.

Near the end of the 1990s—on May 1, 1999—American climber Conrad Anker discovered a body high on the mountain, clad in old-fashioned clothing. Stunned, he realized he had found Sandy Irvine. When the other team members arrived, they discovered a name tag sewn to the collar of one of the shirts: G. MALLORY. So convinced were they that this was Irvine that teammate Jake Norton commented, "That's weird. Why would Irvine be wearing Mallory's shirt?" Anker had found the body of George Mallory.

The following year, Slovenian skier Davo Karničar made the first ski descent of the mountain, taking less than five hours from the summit to Base Camp.

Sherpa Babu Chiri rang in the new millennium when he raced up the normal route in just 16 hours. A ten-time veteran of Everest's summit, including one night's bivouac (planned, and without supplemental oxygen) on the summit in

FRENCH SNOWBOARDER MARCO Siffredi (opposite top) gets set to descend Everest on a snowboard on September 9, 2002. He had successfully snowboarded down the Norton, or Great, Couloir the year before. He returned to attempt the more challenging Hornbein Couloir but would never be seen again.

ERIK WEIHENMAYER (opposite bottom) makes the first blind ascent on May 25, 2001. Close friends led him to the summit. Everest beckons for all kinds of adventure sports, including tandem paragliding from the summit (below). Babu Sunuwar, pilot, and Lakpa Tshering Sherpa descended in just 42 minutes, landing in Namche Bazar.

AS THE SUN rises, climbers push up the final summit ridge before reaching 29,035 feet (8,850 meters).

1999, Babu Chiri died a year later when he fell into a crevasse at Camp II in the Western Cwm while out on an innocent jaunt to take photographs.

French snowboarder Marco Siffredi made the first successful snowboard descent, on the Tibetan side of the mountain the following year. He was killed one year later while trying to snowboard the Hornbein Couloir.

The same year, Erik Weihenmayer defied his naysayers when he became the first blind man to climb the mountain. Despite what some might classify a "stunt," he brought hope to disabled people numbering in the tens of thousands around the world.

Three years later, a Russian team sieged a line to the left of the Japanese Couloir on the North Side, fixing more than 3,000 meters of rope in the process, most of which remained on the mountain after they summited and left.

# THE IMPACT OF EVEREST

Climbing Everest remains for most a life-changing experience. Whether they are guided clients or trailblazing explorers, many climbers have seen their life stories

## THEN & NOW / BACKPACKS

| 1963 | 2012 |
|---|---|

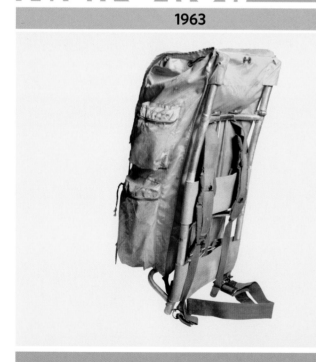

Older, heavier, aluminum-frame packs fastened rigidly to the back, making climbs more effort-intensive. Today's mountain climbers know the comfort of The North Face's lightweight, indestructible backpacks, built with an internal frame that allows more freedom of movement and more storage space.

# "THE MOUNTAIN REMAINS UNKNOWN AS BEFORE, BUT THOSE WHO CLIMBED IT KNOW THEMSELVES A LITTLE BETTER, FOR THE CLIMBING."

—WILFRID NOYCE, ENGLISH MOUNTAINEER, MEMBER OF THE 1953 BRITISH EXPEDITION

dramatically affected by Everest. Ed Hillary was convinced that nobody's life had changed as dramatically as that of Tenzing Norgay's after his ascent of the mountain. In truth, the fame and politicking and posturing humiliated him: "Everest did not matter. Only politics mattered. And I was ashamed." His Everest triumph became an albatross for Tenzing, who spent his later years as a tragically lonely figure.

Fifty years after his climb, the doubts that Tom Hornbein felt on the summit began to make some sense: "Dreams are the beginning, and doubt simply a catalyst to creativity," he said. "My life is still rich with adventure and its attendant uncertainties. Precious are those people with whom I share it." Many of those precious relationships came to Hornbein *because* of Everest.

Reinhold Messner treasured his two Everest climbs: the first supplemental-oxygen-free ascent with Peter Habeler and the lonely solo of the North Face. But his anger over the subsequent developments on Everest eventually overshadowed these feelings. He scoffed with anger that Everest, by 2012, had become a "Disneyland." Instead he turned to smaller peaks. "When I was a small child, I began on small mountains," he said. "Now, as I am getting older, the small peaks are getting bigger. If I am lucky, some day I will end on a small peak."

Everest in winter was Krzysztof Wielicki's first Himalayan climb. It wasn't his last. Like Messner, he climbed all 14 of the 8,000-meter peaks. But for Wielicki, Everest remained his best memory, not because it was highest, but because of the atmosphere, the team effort, and the strong leadership from Andrzej Zawada.

For Wielicki, Everest cemented a deep and meaningful friendship with the man who helped shape his dreams of winter climbing on the highest mountains.

Everest was Alison Hargreaves's last Himalayan peak from which she would return. Just three months later, she was blown off the slopes of K2 after summiting. Almost as violent as her death was the reaction to it, as many vilified her—a mother—for risking so much. The woman whose first words on the top of Everest were for her son and daughter was maligned for not loving them as much as her own ambition.

Friendship, shame, anger, fame, joy. Despite the disparity in the aftermath of their Everest ascents, for each of these climbers the mountain gave a unique, exalted feeling of reaching the top of the world.

Legendary Russian climber Anatoli Boukreev expresses that rapture well: "Mountains are not Stadiums where I satisfy my ambition to achieve, they are the cathedrals where I practice my religion . . . I go to them as humans go to worship. From their lofty summits I view my past, dream of the future and, with an unusual acuity, am allowed to experience the present moment . . . my vision cleared, my strength renewed. In the mountains I celebrate creation. On each journey I am reborn."

**TRIUMPHANT EDMUND HILLARY** (opposite, at right), Tenzing Norgay (opposite center), and John Hunt, a British army officer and the leader of the successful 1953 Mount Everest expedition, arrive at London's Heathrow Airport. After Hillary and Norgay showed the world it could be done, Everest has continued to beckon climbers to test their wills on its icy slopes, as seen from Tibet (below).

# CHAPTER 6

# THE AGONIES

Bruce D. Johnson

American climber Sam Elias
tests his oxygen tank at
Base Camp.

OF EVEREST

# WHY DOES A SCIENTIST TRANSPORT HUNDREDS OF POUNDS OF EQUIPMENT ON THE BACKS OF YAKS AND PORTERS TO SET UP A RESEARCH LABORATORY ON MT. EVEREST?

Why does a scientist transport hundreds of pounds of equipment on the backs of yaks and porters to set up a research laboratory on Mount Everest? While climbers may seek to test the limits of their bodies, scientists are intrigued by the adaptive limits of the human body—in short, by how the climber does it. Humans are capable of a remarkable degree of plasticity, and Everest may be the greatest physiological challenge that exists on our planet, with the severe reduction in available oxygen, the cold, dry conditions, sleepless nights, prolonged time needed for acclimation, and the extreme exertion such a climb requires. To address the challenges of high altitude, to learn how the human body adapts to extreme hypoxia (lack of oxygen), and to find out what happens when the body doesn't adapt, we at Mayo Clinic took the opportunity to partner with National Geographic and The North Face to observe and measure as never before what happens in people's bodies when they climb Everest.

## MEDICAL RESEARCH ON EVEREST

Mount Everest has an ongoing rich tradition of medical research. In any given season, several groups may perform field studies to explore how the body responds

MEMBERS OF THE 2012 Legacy Climb (above) do push-ups and dips to avoid the deconditioning and potential muscle loss that comes with staying at high altitude for an extended time. It's been 50 years since American Tom Hornbein wore this oxygen mask (below) during his ascent of the West Ridge.

**EVEREST'S EXTREME ENVIRONMENT** can cause extreme responses in the human body. Better understanding of how our bodies adjust under these conditions leads to better treatment—and better equipment.

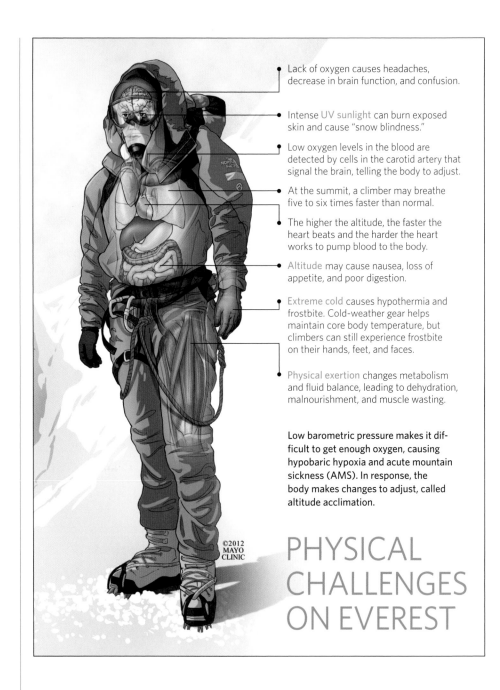

Lack of oxygen causes headaches, decrease in brain function, and confusion.

Intense UV sunlight can burn exposed skin and cause "snow blindness."

Low oxygen levels in the blood are detected by cells in the carotid artery that signal the brain, telling the body to adjust.

At the summit, a climber may breathe five to six times faster than normal.

The higher the altitude, the faster the heart beats and the harder the heart works to pump blood to the body.

Altitude may cause nausea, loss of appetite, and poor digestion.

Extreme cold causes hypothermia and frostbite. Cold-weather gear helps maintain core body temperature, but climbers can still experience frostbite on their hands, feet, and faces.

Physical exertion changes metabolism and fluid balance, leading to dehydration, malnourishment, and muscle wasting.

Low barometric pressure makes it difficult to get enough oxygen, causing hypobaric hypoxia and acute mountain sickness (AMS). In response, the body makes changes to adjust, called altitude acclimation.

©2012 MAYO CLINIC

# PHYSICAL CHALLENGES ON EVEREST

to the extreme conditions on Everest. The adventuresome scientist finds this type of work intriguing, but it also has practical applications in other areas of medicine. For example, knowledge gained on Everest can be used to assist medical personnel during the deployment of troops or workers to cold, high, and dry environments or equally extreme conditions. At the U.S. South Pole Station, for example, up to 700 workers are rapidly exposed to high-altitude conditions each year. Others serve in oil-industry postings in extreme northern latitudes or work in the numerous observatories located at high altitudes. And of course there's the general outdoor sports/tourist industry.

But there are also applications in more everyday settings, such as the hospitals and clinics that treat patients with congestive heart failure, sleep disorders, lung disease, and other conditions similar to the hypoxia, or lack of oxygen, experienced by climbers and others at high altitudes.

## HIGH-ALTITUDE CHALLENGES

What is it about higher altitudes that challenges the body?

In the 1600s, on the heels of work by Galileo, an Italian physicist, Evangelista Torricelli (1608–1647), determined that the air above us has weight. Around this same time physicist Robert Boyle (1627–1691) and others discovered that the weight of air (or air pressure) determines the level of oxygen in our lungs and subsequently in our bloodstream. This explains the challenge of surviving at high altitudes: Air pressure drives oxygen into our blood, and the agony felt by climbers comes from a severe lack of air pressure.

Before we can explore how altitude affects the body, we must first understand how the body processes and uses oxygen on the Earth's surface. We breathe oxygen into our lungs, and from there the oxygen enters the bloodstream, traveling to body organs that require oxygen to operate. Hemoglobin is the metalloprotein in red blood cells responsible for transporting oxygen throughout the body. Blood that has run through the body returns to the lungs for a fresh infusion of oxygenated air, which means that on average the amount of oxygen in the lungs is greater than the amount of oxygen in the arteries.

Oxygen saturation values (one measure of the amount of oxygen in the

# "WITH THE MASK ON I FELT AS IF I WERE WEARING CYRANO'S NOSE."

—BARRY C. BISHOP, *NATIONAL GEOGRAPHIC*, 1963

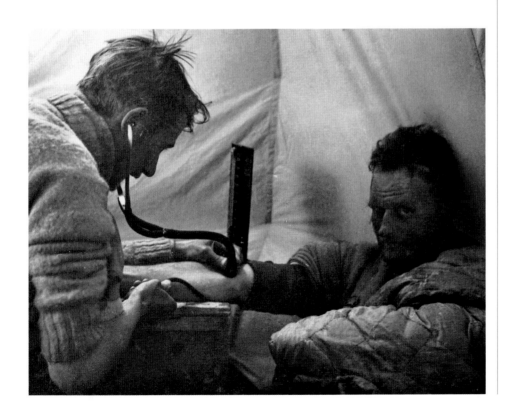

**DR. CHARLES WARREN** takes climber Eric Shipton's blood pressure during a 1938 British Everest expedition. Physicians then were learning what we know now: Individual responses to Everest's harsh environment vary dramatically.

**SHERPAS SHUTTLE LOADS** of supplies in the Western Cwm. A quest for the summit usually takes six to eight weeks of acclimation, including a gradual climb to Base Camp and multiple trial-and-error advances to higher altitudes to see how the body adjusts.

bloodstream) normally range between 95 and 100 percent. At sea level it is difficult to hold one's breath long enough to cause a significant drop in these values. When values fall below 85 to 88 percent, as can happen among patients with chronic lung disease, it often becomes necessary to use portable or home oxygen in order to maintain a normal quality of life.

Almost everywhere on the surface of the Earth, the amount of oxygen in the atmosphere remains stable at 21 percent. In fact, from sea level up to the limits of the Earth's atmosphere (62 miles, or 100 kilometers, above the surface), oxygen levels in the air remain nearly constant. What changes is air pressure, which drops proportionally as altitude increases. The atmospheric pressure at Everest Base Camp is about half that at sea level; at the summit, it's one-third that at sea level. In other words, each breath of air taken in at high altitude contains less oxygen than at sea level. As a result, oxygen levels in the lungs and in the blood drop in parallel with the atmospheric drop in pressure. Normal oxygen saturation levels at 10,000 feet (3,000 meters, just above the starting point for most Everest expeditions) are between 88 and 90 percent. During sleep, when breathing slows, these levels may fall below 84 percent—lower than the point at which many lung-disease patients use supplemental oxygen.

The balance between oxygen levels in the lungs and arteries is especially important when climbing at high altitudes. Physical exertion increases the speed at which red blood cells travel through the lungs and thus reduces the time blood spends getting replenished with oxygen in the lungs. The combination of exercise and high altitude doubly stresses the body, more and more the higher a person climbs. (I should point out here that any given individual's ability to adapt to high altitudes can vary dramatically and depend on a variety of factors including genetics, overall health, and altitude of residence, as well as a number of situational factors such as hydration levels.)

**BRITISH CLIMBER STEPHEN** Venables returned 20 pounds lighter after his ascent of the Kangshung Face, the east side of Everest, without oxygen in 1988. Extended time at altitude can lead to rapid weight loss, which includes the loss of muscle.

## MOUNTAIN DANCES

The body is equipped to handle modest changes in altitude. For example, changes in air pressure at altitudes up to 8,000 to 10,000 feet (2,500 to 3,000 meters) only mildly reduce the oxygen saturation in the body, thanks to the unique properties of hemoglobin. But at greater heights, the lack of oxygen can have serious—and even fatal—consequences. Without the benefits of the acclimation process—in other words, if a person moves rapidly from sea level to an extreme altitude—most, if not all, individuals would become sick above 15,000 feet. They would likely be conscious for only 20 to 30 minutes at 18,000 feet and only for 1 to 2 minutes at the summit of Mount Everest.

In order to avoid the worst effects of altitude sickness, most climbers who plan to summit Everest spend six to eight weeks in a laborious dance with the mountain and their health. This dance of acclimation includes a gradual ascent to Base Camp followed by several rotations to higher and higher altitudes, the climbers slowly pushing the

**CONRAD ANKER'S EXHALED** air is monitored at Base Camp by Bruce Johnson. Breathing increases dramatically at high elevations, causing carbon dioxide levels in the blood to fall. Acclimation allows this drop to occur more aggressively, important in the adaptation process.

"hypoxic stress" in hopes of stimulating the body to adapt to higher altitudes. While the body becomes more able to accommodate the extreme altitudes, sustained time spent at altitudes as high as Everest Base Camp and above can cause muscle deconditioning and muscle loss. Thus a balance must be negotiated between acclimating to heights and losing muscle condition. Some climbers even go down to lower altitudes for brief periods of time to gain back some muscle strength. The protocol for acclimation has primarily been developed by trial and error, taking advantage of the impressive adaptive abilities of the human body, particularly for short windows of time.

## IMMEDIATE DEFENSE

How does the body respond to reduced oxygen levels in the blood and a reduced amount of oxygen delivered to the tissues? The amount of oxygen delivered in the bloodstream to body tissues is a product of several factors: the number of red blood cells (hemoglobin levels); how saturated the hemoglobin molecules are with oxygen molecules (arterial oxygen saturation); and the output of the heart (the heart rate and the volume pumped per beat of the heart).

The first response to lowered oxygen levels occurs almost immediately. Breathing speeds up so that the amount of air inhaled increases. This boosts the average oxygen levels in the lungs and subsequently in the blood. Heart rate also rises, which pushes more blood into the body and boosts the delivery of oxygen to the body's tissues. In addition, soon after a person reaches high altitude, urine production increases, and some fluid may shift from within blood vessels to outside. This reduces the water in the bloodstream and makes the red blood cells more concentrated. That in turn increases the amount of oxygen transported for a given volume of blood. The reduction of liquid in the bloodstream, however, results in a drop in the volume pumped per beat of the heart. (A side effect of this process is dehydration, which can increase the likelihood of acute mountain sickness, or AMS, characterized by headache, nausea, and fatigue. Thus drinking large amounts of water is essential at high altitudes.)

The hemoglobin molecule in red blood cells works differently depending on many factors: pH level, carbon dioxide level, temperature of the blood, for example. These properties of hemoglobin play an important part in how the body adapts to high altitudes. For example, as breathing increases at high altitude,

## THEN & NOW BOOTS

| 1963 | 2012 |
|------|------|

Reindeer boots of yore still hold charm in their appearance, but mountain climbing has become high-tech adventure. Lacing these heavy, slow-to-dry boots with hands encased in thick gloves was one of many challenges. Today's designs include foam insulation, removable liners for fast drying, a heat-reflective aluminum lining for extra warmth, and a durable front zipper.

chortens, at the mountain pass of Thukla on the way to Base Camp honor climbers and Sherpas who have died on Everest. Covered in prayer flags, the memorials stand as a sobering reminder that not everyone makes it back from the mountain.

the amount of carbon dioxide in the blood falls, reaching as low as one quarter of normal values by the time the climber reaches the summit of Everest. This causes blood to become more alkaline (making the pH level higher) and changes hemoglobin's ability to bind oxygen molecules. This enables red blood cells at a high altitude to attract and bind oxygen better, despite the lower oxygen pressure in the lungs. The same factors can also influence the release of oxygen to tissues.

Other responses to low oxygen levels include a constriction of blood vessels within the lungs, which raises the pressure levels in the lungs and on the right side of the heart. This effect can be almost immediate, but the level may continue to rise over time and with higher altitudes. Although high blood pressure in the lungs can have negative consequences over a long period of time, in the short term it may redistribute blood to more regions of the lungs, increasing the total surface area for the transfer of gases and moving oxygen into the bloodstream and carbon dioxide out to the lungs.

## LONGER-TERM ADAPTATION

As the "stress" of high altitude continues, the body's many low-oxygen-sensing genes, proteins, and cells regulate how the body adapts. The stimulation of one

# EFFECTS OF HYPOXIA

Hypoxia: A condition in which the body as a whole or a region of the body is deprived of oxygen supply.

Low oxygen pressure at high altitude.

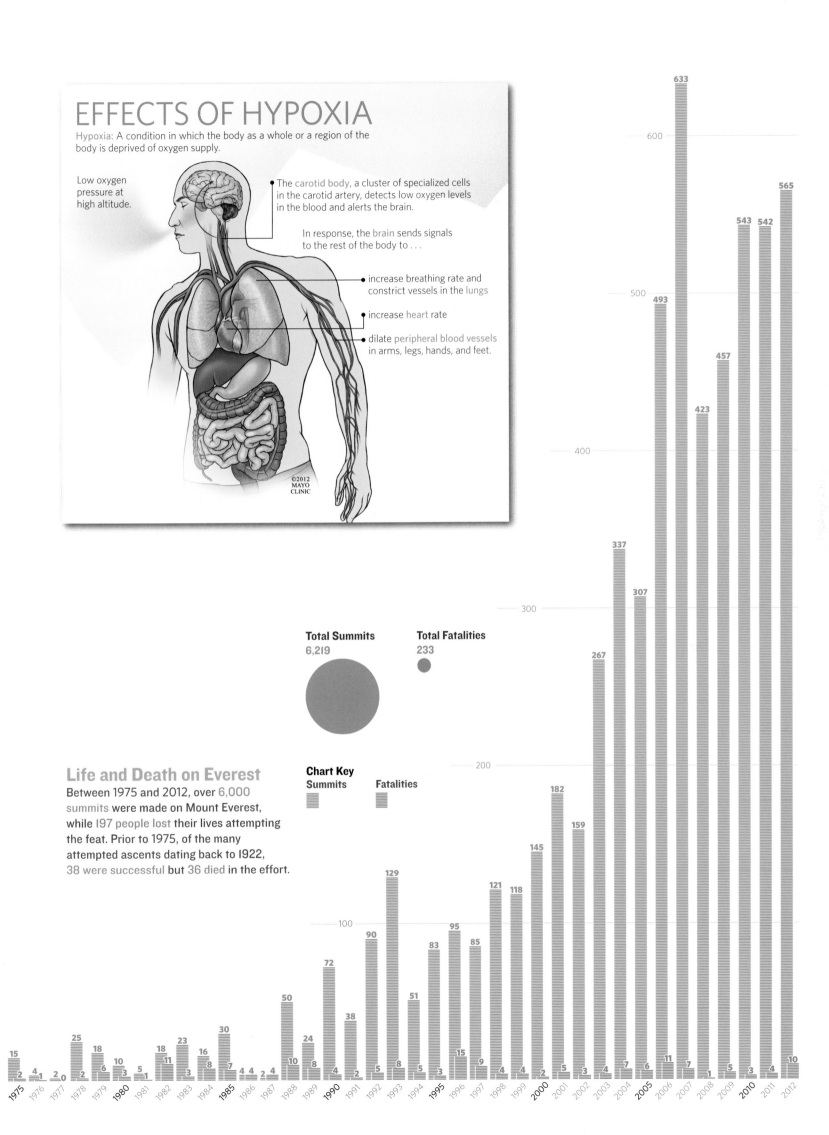

The carotid body, a cluster of specialized cells in the carotid artery, detects low oxygen levels in the blood and alerts the brain.

In response, the brain sends signals to the rest of the body to . . .

increase breathing rate and constrict vessels in the lungs

increase heart rate

dilate peripheral blood vessels in arms, legs, hands, and feet.

©2012 MAYO CLINIC

**Total Summits**
6,219

**Total Fatalities**
233

**Chart Key**
Summits     Fatalities

## Life and Death on Everest

Between 1975 and 2012, over 6,000 summits were made on Mount Everest, while 197 people lost their lives attempting the feat. Prior to 1975, of the many attempted ascents dating back to 1922, 38 were successful but 36 died in the effort.

| Year | Summits | Fatalities |
|------|---------|-----------|
| 1975 | 15 | 2 |
| 1976 | 4 | 1 |
| 1977 | 2 | 0 |
| 1978 | 25 | 2 |
| 1979 | 18 | 6 |
| 1980 | 10 | 3 |
| 1981 | 1 | |
| 1982 | 18 | 11 |
| 1983 | 23 | 8 |
| 1984 | 16 | 3 |
| 1985 | 30 | 7 |
| 1986 | 4 | 4 |
| 1987 | 2 | 4 |
| 1988 | 50 | 10 |
| 1989 | 24 | 8 |
| 1990 | 72 | 4 |
| 1991 | 38 | 2 |
| 1992 | 90 | 5 |
| 1993 | 129 | 8 |
| 1994 | 51 | 5 |
| 1995 | 83 | 3 |
| 1996 | 95 | 15 |
| 1997 | 85 | 9 |
| 1998 | 121 | 4 |
| 1999 | 118 | 4 |
| 2000 | 145 | 2 |
| 2001 | 182 | 5 |
| 2002 | 159 | 3 |
| 2003 | 267 | 7 |
| 2004 | 337 | 6 |
| 2005 | 307 | 11 |
| 2006 | 493 | 7 |
| 2007 | 633 | 1 |
| 2008 | 423 | 5 |
| 2009 | 457 | 3 |
| 2010 | 543 | 4 |
| 2011 | 542 | |
| 2012 | 565 | 10 |

# "ALTHOUGH WE ARE TOTALLY DEBILITATED, OUR SUCCESS BREEDS EUPHORIA."

–BARRY C. BISHOP, *NATIONAL GEOGRAPHIC*, 1963

of these proteins (called hypoxia inducible factor 1 alpha, or HIF-1 $\alpha$) sets off a cascade of changes throughout the body, including the stimulation of a hormone (erythropoietin) that controls red blood cell production. This then increases the blood's carrying capacity of oxygen.

Other longer-term changes include the increased ability of the kidneys to excrete bicarbonate, which buffers against swings of pH levels in the blood. This critical adaptation lowers the pH level of the blood and allows the gradual ramping up of breathing to higher and higher levels over several weeks at altitude. In addition, there is an increase in oxygen-carrying proteins in muscle called myoglobin—a relative of hemoglobin. The ratio of capillaries (branching blood vessels) to muscle fibers may increase and improve ability to supply oxygen to muscle. (Muscles are made up of many smaller fibers.) Many climbers

A SHERPA WOMAN in the Khumbu Valley tends her field. Research suggests that centuries of high-altitude life have shaped the Sherpa predisposition to breathe deeply without suffering from hypoxia.

I suppose I had it coming to me. I was the victim of my own ambition, my own weakness, the cold, the altitude . . . I paid the price for pushing on late in the day. There were moments that afternoon when I thought I had realized my worst terror—driving myself to the top of the world then finding I had no reserves left to get back down. After one frantic, blind glissade down the steep flank of the South Summit, I collapsed in a wild, gasping, fish-out-of-water panic, dug a hole in the snow, and just sat shivering violently, feeling very sorry for myself.

But it was a precarious perch, and I knew I had to get a grip and stop being a victim. At dusk I climbed down to an easier section of the ridge. As darkness enclosed the mountain I switched on my headlamp and struggled to make sense of half-remembered rocks I had passed on the way up, ten hours earlier. (Not knowing the correct normal route we had climbed a steeper, more direct line, gaining the Southeast Ridge some way above "The Balcony.") In the end I decided it would be safer to wait for the dawn. So I settled down for the long shiver.

At first I sat on a rock. Rocks are warmer—I knew that. But this one sloped and I longed, no we longed—my imaginary older companion and I—to lie down flat. And I knew anyway that it's better for your circulation if you stretch your legs horizontally. So we got up and started digging; but we were so tired, so feeble, so desperate to lie down, that we only managed a five-foot-long ledge in the snow before settling down with knees bent out over the Kangshung Face.

By now it must have been dark for at least an hour and I still had seven hours incarceration before the release of daylight. I later calculated the altitude at about 8,550 metres above sea level. Night temperature in May: somewhere between -30° and -40° centigrade. Chances of survival? Well, actually, not too bad. The afternoon blizzard had subsided and the air was still . . .

Although I never planned, or wanted, to bivouac on Everest, the grim possibility had always been lurking there, unspoken. I had thought very carefully about all my layers of clothing. And I now took care to pad hips and head with spare mittens—token wads of extra insulation between my bones and the insidious creeping cold. I also plunged my long ice axe into the snow as a fence post to stop me rolling over during the night.

There is a myth that if you fall asleep in these situations you never wake up . . . I think at one point I did

Stephen Venables on Everest, 1983

actually slumber for a while, and I didn't die. Mostly though I was in semi-conscious limbo, drifting through a fog of half-rational associations . . . Occasionally my rational consciousness would jolt me awake and banish the phantoms. I forced myself to chew some almond Hershey bar—must be some calories there. I sucked the last semi-frozen trickle from my water bottle. I thought about removing crampons and boots to massage wooden toes, but decided it was too risky—better to concentrate on looking after my hands. And I shivered. But it didn't really feel any worse than my worst alpine winter bivouacs 14,000 feet lower: I was anaesthetised by altitude. And in any case, I was no longer a victim: I was just waiting patiently for the dawn.

**—STEPHEN VENABLES  In 1988 mountaineer, writer, and inspirational speaker Stephen Venables became the first Briton to summit Mount Everest without supplementary oxygen.**

loads at high elevation in the Khumbu Valley. Studies have shown that Sherpas rarely suffer from acute or chronic mountain sickness.

experience a rapid loss of weight and muscle at extreme altitude. (The melting away of muscle without a change in the number of capillaries may explain the increased capillary-to-fiber ratios.)

## HOW HIGH FOR HOW LONG?

What are the true limits to high-altitude exposure? While many call the process preparing for an Everest summit "acclimation," others would say the body gets better at "tolerating" the extremes of hypoxia (low oxygen). That tolerance continually remains at a delicate threshold, particularly at the elevations above Everest Base Camp. That's why over the course of history, few long-standing communities have survived at extreme altitudes. In fact, no permanent communities on the globe exist above 17,000 feet. (For comparison, Everest Base Camp is at about 17,500 feet.) Most of the communities even close to that altitude grew up around industry (mining towns, railway communities, or observatories) and struggle with ongoing problems related to the extreme altitude. The highest permanent mining community is La Rinconada in southern Peru, which extends from altitudes of 4,900 to 5,200 meters (16,000 to 17,000 feet). In

Tibet several villages are located at similar elevations, but most settlements are near or below 15,000 feet, which seems to be a threshold for maintaining health and reproduction. Even the highest Sherpa villages are lower than 15,000 feet, or 4,600 meters, high. (Dingboche is at 4,400 meters and Pheriche 4,240 meters.)

The Tibetan and Andean people who live in the highest plateau regions of the world provide an interesting comparison of how humans can adapt to extreme elevations. Although we do not know how long each of these groups has lived at high altitude, the consensus is that the Tibetans settled their plateau communities first and thus have had a greater chance for natural selection and adaptation to high altitudes. Both populations have adapted to their environment and appear to have relatively normal energy expenditures at rest and relatively adequate exercise capacities, even at altitudes up to about 4,000 meters. This contrasts greatly with most healthy adults living at sea level, whose exercise capacity falls by 20 to 30 percent at these altitudes. Although exercise tolerance can increase over time at high altitude, the Tibetan and Andean inhabitants show relatively comprehensive adaptation to high altitudes.

The two groups may have achieved this acclimatization by different means, however. For example, the Tibetans appear to breathe more than the Andeans and have more blood flow through the lungs, despite lower average hemoglobin levels and oxygen saturation values. The Tibetans appear less susceptible than Andeans to two conditions associated with living at high altitudes: chronic mountain sickness (CMS), in which the number of red blood cells increases to the point where the blood gets sludgy and difficult for the heart to pump it throughout the body, and high blood pressure in the lungs, a risk factor for CMS. Lack of oxygen may contribute over time to the selection of genes that help the acclimation process. Scientists do not know whether the Tibetans' apparently more complete adaptation is due to the number of generations they have lived at high altitude. Discovering the genetic basis for these adaptations may help

THOUSANDS OF MILES from Everest in the Andes of Peru, a woman harvests potatoes on a high-altitude hillside. Like Sherpas, Andeans have adapted to living at an extreme elevation, but chronic mountain condition is more common, and they appear more susceptible to high blood pressure in the lungs.

CLIMBERS WALK PAST the body of Canadian Shriya Shah-Klorfine, who died on her descent from the summit in 2012. The 33-year-old, who was relatively inexperienced, was reportedly told by her guides to turn around before she reached the top, but she refused. She collapsed in the "death zone," the icy slope above the last camp.

find better ways to help the body tolerate low oxygen levels—important for many patients—or may help identify those who will either do well or face high risk at high altitudes.

# SUPPLEMENTAL OXYGEN

The use of oxygen on Mount Everest remains controversial. For years many thought that climbers could not survive without supplemental oxygen, but eventually Reinhold Messner and Peter Habeler proved otherwise in 1978. Of the more than 3,500 people to reach the summit of Everest, roughly 5 percent have done so without oxygen. Still, a climber is almost two times more likely to die while summiting Everest without using oxygen.

Dr. Thomas Hornbein, a member of the original U.S. expedition to summit Everest in 1963, estimated the influence of supplemental oxygen. Mayo Clinic researchers recently modeled these results using data from the 2012 expedition and a simulated ascent of Everest that took place in 1985. In that study, a team of 27 investigators "climbed" for 40 days as if to the summit in an altitude chamber. Results show that a climber using supplemental oxygen will have a hypoxic stress level at the summit that is lower than the one experienced at Base Camp at rest without oxygen. With oxygen, a climber at Camp IV trying to sleep actually maintained oxygen saturation levels in the mid to upper 90s, close to sea-level values. Thus supplemental oxygen makes a big difference, although some might say that six weeks of acclimating to extreme hypoxia results in muscles less able to use the oxygen being delivered.

# THE ULTIMATE LABORATORY

Attempting to mimic extreme conditions in the laboratory is difficult, if not impossible. The field is the ultimate laboratory where human physiology and environment come together. Few atmospheric chambers can mimic true high-altitude conditions, complete with unpredictable temperature fluctuations and humidity levels as well as human factors such as dehydration, physical activity, and psychological stress. Field work on Everest complements the highly controlled laboratory conditions. The less predictable conditions in the field can lead to observations and findings that had not been considered or explored.

**1963 CLIMBER BARRY** Bishop permanently lost toes during the historic climb. Rates of frostbite (above) have been reduced over the years thanks to improvements in equipment.

**MARK JENKINS (BELOW)** learns to use an oxygen mask system. At Base Camp there is only half the oxygen available that there is at sea level; on the summit, 33 percent. Five percent of people who have summited Everest have done so without supplemental oxygen.

## 2012 STUDIES AND BEYOND

Fifty years after the first American ascent of Everest, the names and faces have changed but the challenge of the climb and the passion for unlocking the scientific secrets of high-altitude physiology remain alive and well. Fifty years ago, the climbing "team" consisted of 19 members, including five with master's degrees, five with Ph.D.'s, three medical doctors, and a group of experienced climbers and their support staff. For the studies of the medical aspects of high-altitude exposure, primarily cognitive and psychological measures, the study subjects were

MEMBERS OF THE first American team to summit Everest, in 1963, are weighed in. Part of the expedition's mission was to study the effects of high-altitude exposure on the body.

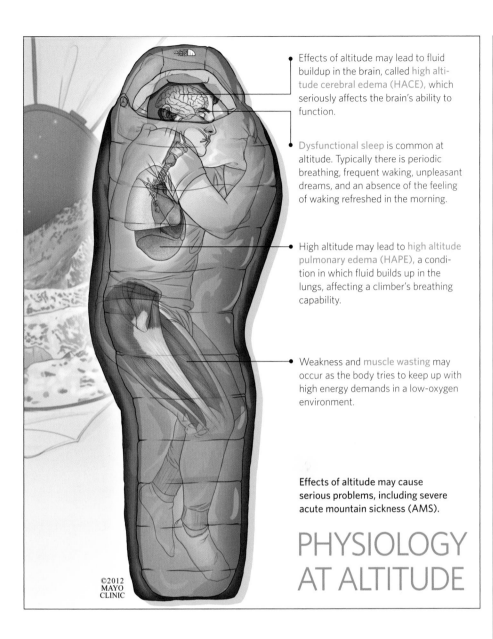

Effects of altitude may lead to fluid buildup in the brain, called high altitude cerebral edema (HACE), which seriously affects the brain's ability to function.

Dysfunctional sleep is common at altitude. Typically there is periodic breathing, frequent waking, unpleasant dreams, and an absence of the feeling of waking refreshed in the morning.

High altitude may lead to high altitude pulmonary edema (HAPE), a condition in which fluid builds up in the lungs, affecting a climber's breathing capability.

Weakness and muscle wasting may occur as the body tries to keep up with high energy demands in a low-oxygen environment.

Effects of altitude may cause serious problems, including severe acute mountain sickness (AMS).

## PHYSIOLOGY AT ALTITUDE

©2012 MAYO CLINIC

essentially members of the team. Support then as now was an ongoing challenge, but the National Geographic Society, National Science Foundation, U.S. Air Force, Office of Naval Research, the State Department's Bureau of Educational Exchange and Cultural Affairs, the U.S. Army Quartermaster Corps, and NASA all provided significant funding for the 1963 climb.

This year's team consisted of a similar number of individuals, along with multiple funding sources, including substantial support from the National Geographic Society and The North Face company. Team members included Ph.D. physiologists and geologists, physicians with expertise in critical care and flight medicine (USAF), a research associate, carefully selected North Face and National Geographic athletes, athlete-photographers, videographers, bloggers, product-innovation experts with The North Face company, and a National Outdoor Leadership School senior field instructor.

> ## "I HAVE NOT CONQUERED EVEREST. IT HAS MERELY TOLERATED ME."
>
> —PETER HABELER, AUSTRIAN MOUNTAINEER WHO, WITH REINHOLD MESSNER, MADE THE FIRST ASCENT OF EVEREST WITHOUT SUPPLEMENTAL OXYGEN, 1978

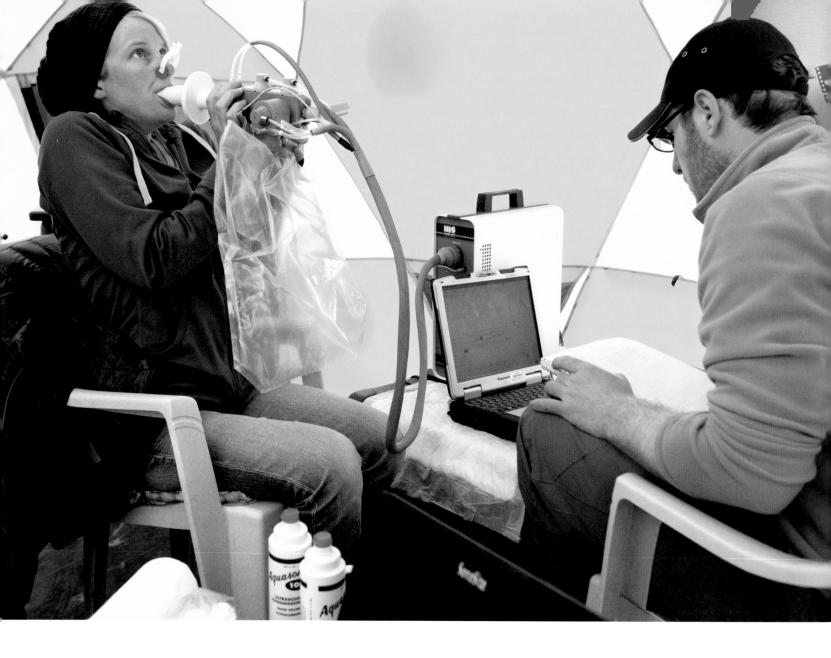

**ELITE CLIMBER EMILY**
Harrington, a member of the 2012 North Face/National Geographic expedition team, takes a lung-capacity test (above). Part of the team's mission was to study the limits of performance at high elevation. All team members—such as photographer Cory Richards (opposite)—underwent physiological testing prior to departing for Everest.

**HARRINGTON STRAPS ON**
her crampons before making her way through the Khumbu Icefall (following pages). Climbers often perform this climb in the middle of the night, when the ice is most stable.

## THE HIGH-REACHING GOALS

The North Face company develops products based on feedback from the top athletes around the world, people who push the envelope on human performance in extreme and unlikely environments. This focus on human performance in extreme environments like Everest became the impetus for the North Face–Mayo Clinic relationship, since the Mayo Clinic also has an extensive history of such studies. Mayo researchers played a critical role in the early understanding of flight physiology and health countermeasures for pilots. While researchers can learn a lot from studying chronic disease (one end of the health spectrum), our belief is that there is also much to be learned from studying the exceedingly fit athletes who push the other end of the health spectrum.

The Mayo Clinic researchers had several goals for the 2012 Everest expedition. One was to better understand the limits of human performance at extreme altitudes. They also wanted to determine the relationship between altered sleep physiology (amount of REM sleep and/or the decrease in of blood-oxygen levels at night) and altitude illness. They studied lung-fluid regulation and muscle loss, and they collaborated to evaluate the impact of equipment and clothing on sleep physiology and performance in athletes. The studies on oxygen deprivation complemented ongoing clinical studies at the Mayo Clinic with patients. Climbers wore devices that were light and unobtrusive even during difficult ascents, which allowed the continuous monitoring of participants 24 hours a day and gave us a more dynamic and comprehensive view of the body under extreme conditions. Researchers collected an estimated 27 billion data points over the course of the expedition.

## WHAT WE LEARNED

It is fascinating to realize how much what we learned about adaptive physiology in extreme altitudes applies to the treatment of several chronic diseases. In particular, our laboratory has been interested in patients who suffer from chronic heart failure, a condition in which the pumping action of the heart is less than ideal.

## THEN & NOW — SLEEPING BAGS

| 1963 | 2012 |
|------|------|

The down sleeping bag—a highly coveted innovation of the 1960s—has been improved to better accommodate mountaineers. The North Face's waterproof Inferno bags used on the 2012 Everest expedition include side-block baffles and a vaulted foot box for extra protection.

The American Heart Association estimates that in 2010, some six million patients in the United States experienced heart failure, at a cost of nearly $40 billion. Roughly 700,000 new cases are diagnosed annually. Heart failure is the most common diagnosis in Medicare beneficiaries, and these types of patients have very high admission and readmission rates. The causes of this complex condition are cumulative and broad in origin (e.g., high blood pressure, heart attack, sleep apnea, alcohol, viruses). Many of the current treatments and medications target the body's response to heart failure rather than having a direct therapeutic influence on the heart.

The parallels of heart failure and high-altitude extremes are therefore interesting. A quote from *National Geographic* field staff writer Mark Jenkins describes what it feels like to climb at extreme altitudes: "Duct-tape two bricks to the bottom of each foot, put a straw in your mouth, and then charge up several flights of stairs—breathing only through the straw, of course. If you find you're sucking wind like a race horse, nostrils flaring, mouth drooling, and your heart is jackhammering in your chest, and your legs are as heavy as bags of concrete—welcome

**AMERICAN JESSE RICKERT** traverses the North Face of Everest en route to the summit. The symptoms of fatigue on Everest mimic those of a person suffering from heart failure: extreme exhaustion, difficult breathing, nausea, and light-headedness.

to high altitude." The most common symptoms at high altitudes include short-ness of breath with exertion, extreme fatigue, trouble sleeping and difficulty breathing, weakness and malaise, peripheral edema, lung congestion, headache, nausea, and dizziness or light-headedness—many of the same symptoms that bring heart-failure patients to their local physician.

As a climber reaches high altitude, a small, complex organ called the carotid body—located near the neck area—senses the change in oxygen in the blood and tells the brain to immediately increase breathing. The carotid body also stimulates the nervous system, like a stress response. That sets off a number of processes within the body, such as an increase in nervous system activity and release of mediators such as adrenaline. Recent tests have shown that the reduc-tion in blood flow after heart failure—caused by a drop in the heart's pumping action—stimulates a response in the carotid body that is similar to what occurs at high altitude, when it senses low blood flow as hypoxia. The similarities also extend to the carotid body's stimulating overbreathing in heart-failure patients and stimulating the nervous system as a stress response.

**A LENTICULAR CLOUD** hovers over the summit of Everest like a halo. Studying the effects of altitude on climb-ers in pursuit of reaching this peak can help us understand conditions and functions of the body at sea level.

**A HELICOPTER FLIES** near the periphery of the upper Khumbu Basin (below). Helicopter rescues were once a last resort on Everest but are now increasingly common, thanks to lightweight craft that can operate well at high altitude.

**A MAYO CLINIC** researcher at Base Camp (opposite) connects climber Kris Erickson to a Mayo Clinic sleep-monitoring device to learn about oxygen intake during sleep and its effects on altitude sickness.

There are more parallels between high-altitude exposure and heart failure. The overbreathing, or hyperventilation, that is key to adaptation at high altitudes negatively affects the brain's blood flow, which causes constriction of the blood vessels to the brain and a decrease in carbon dioxide in the blood. Thus a battle emerges between the carotid body telling the brain to breathe and the brain trying to reduce breathing to improve its own blood flow. This tug-of-war occurs especially at night during periods of cluster breathing, causing episodes that resemble sleep apnea—that is, when a person stops breathing for more than ten seconds. Often the climber at high altitude awakens gasping for air, a condition called periodic breathing at high altitude. Heart-failure patients frequently develop a similar form of breathing called Cheyne-Stokes respiration, which is a period of panting followed by deep breaths. Associated with this condition is central sleep apnea (CSA)—common in heart failure. The periods of apnea at high altitudes not only cause more frequent awakenings and poorer sleep quality but also deeper dips in blood-oxygen levels at night. This likely increases the hypoxia stress response. One of the hypotheses of the Everest project was that these dips

contribute to a greater stress response and higher risk for altitude-related illnesses. Both Cheyne-Stokes breathing and CSA in heart failure are also associated with higher risk of death.

## HIGH ALTITUDES AND HEART FAILURE

High-altitude climbers can also face a problem if the blood vessels in their lungs constrict. This causes a high blood pressure condition in the lungs, known as pulmonary hypertension, that in turn causes increased shortness of breath at high altitude and can lead to water accumulating in the lungs, known as high altitude pulmonary edema (HAPE). HAPE can be a life-threatening condition, and climbers suffering from it may be evacuated from the mountain. HAPE occurs at reported rates that range from about 2 percent to as high as 70 percent, depending on how fast the ascent is made, the final altitude, the altitude at which the climber sleeps, and the method of lung-fluid detection. With relative rapid ascents to 4,500 meters, the rate at which HAPE occurs most likely approaches 12 to 15 percent.

Similarly, heart-failure patients often have lung congestion, including fluid in the lungs. This usually occurs because of a backup of blood and pressure in the lungs from decreased function in the left ventricle.

The way the body regulates fluids in the lungs is a complex subject, but recent studies in climbers may shed some light. Some studies have suggested that high-altitude exposure may cause a small rise in lung fluid in most climbers.

# "YOU DON'T JUST DEAL WITH ADVERSITY. YOU USE IT TO PROPEL YOURSELF FORWARD."

–ERIK WEIHENMAYER, THE FIRST BLIND MAN TO SUMMIT EVEREST

However, in only a few does this actually degrade to the life-threatening condition of HAPE. Release of stress mediators, such as adrenaline and deeper breathing, may enhance fluid removal. Other, recent studies have suggested that a common inhaled medication for patients with chronic lung diseases such as asthma—called a beta agonist—may stimulate the clearance of fluid at high altitudes and reduce the incidence of lung edema. Use of this medication, long thought to have a negative impact on heart-failure patients, may actually stimulate the clearance of fluid for them as well.

While the arterial blood in patients with chronic heart failure has relatively normal oxygen saturation values, the heart's reduced pumping power causes a high loss of oxygen from the supply; and only a low amount feeds the tissues while the blood circulates. In addition, central sleep apnea may cause additional dips in oxygen at night. This can create a condition that resembles hypoxia (oxygen depravation) in sicker heart-failure patients.

Both heart failure and high altitudes cause marked reductions in how much oxygen is delivered to the body's organs and tissues. On the summit of Mount Everest, oxygen delivery to a healthy athlete climber's tissues seems to be similar to that of a relatively sick heart-failure patient at sea level.

## UNSOLVED MYSTERIES

During the recent expedition, as in the 1963 expedition, climbers suffered through the many days and nights of climbing Mount Everest. Most had to overcome illness, sleepless nights, weather delays, emotions, and poor conditions to reach the summit of the highest mountain in the world. As in 1963, the physiology and mental functions of the climbers were recorded, cataloged, and analyzed in light of these challenges. How and why do some make it to the top while others do not? Why do some adapt to lower air pressure while others do not? The answers remain unclear.

But we can take the lessons from Everest and apply them to the larger arena of human adaptation and health. Extreme altitudes provide an important model for studying the acute and chronic effects of oxygen deprivation on the human body. Scientists gain small but important increments in knowledge with each expedition, and these contributions get integrated into the body of medical knowledge and improve patient care. The 2012 Everest expedition involved a new generation of researchers who will carry forward this exciting endeavor to expand the realms of medical care.

**CLIMBER AND PHOTOGRAPHER** Barry Bishop (opposite), on the 1963 American team, is helped down the mountain due to his badly frostbitten feet. He eventually lost all of his toes and the tips of his little fingers.

**BISHOP HAD TO** be carried out by Sherpas (below) for helicopter evacuation. He once said, "Everest is a harsh and hostile immensity. Whoever challenges it declares war. He must mount his assault with the skill and ruthlessness of a military operation. And when the battle ends, the mountain remains unvanquished. There are no true victors, only survivors."

# CHAPTER 7

# ONE SEASON

Mark Jenkins

The sun sets behind the
Khumbu Glacier of Everest
as seen from Camp II in
the Western Cwm. Everest
rises to the right, and
the neighboring peak of
Nuptse rises to the left.

# ON EVEREST

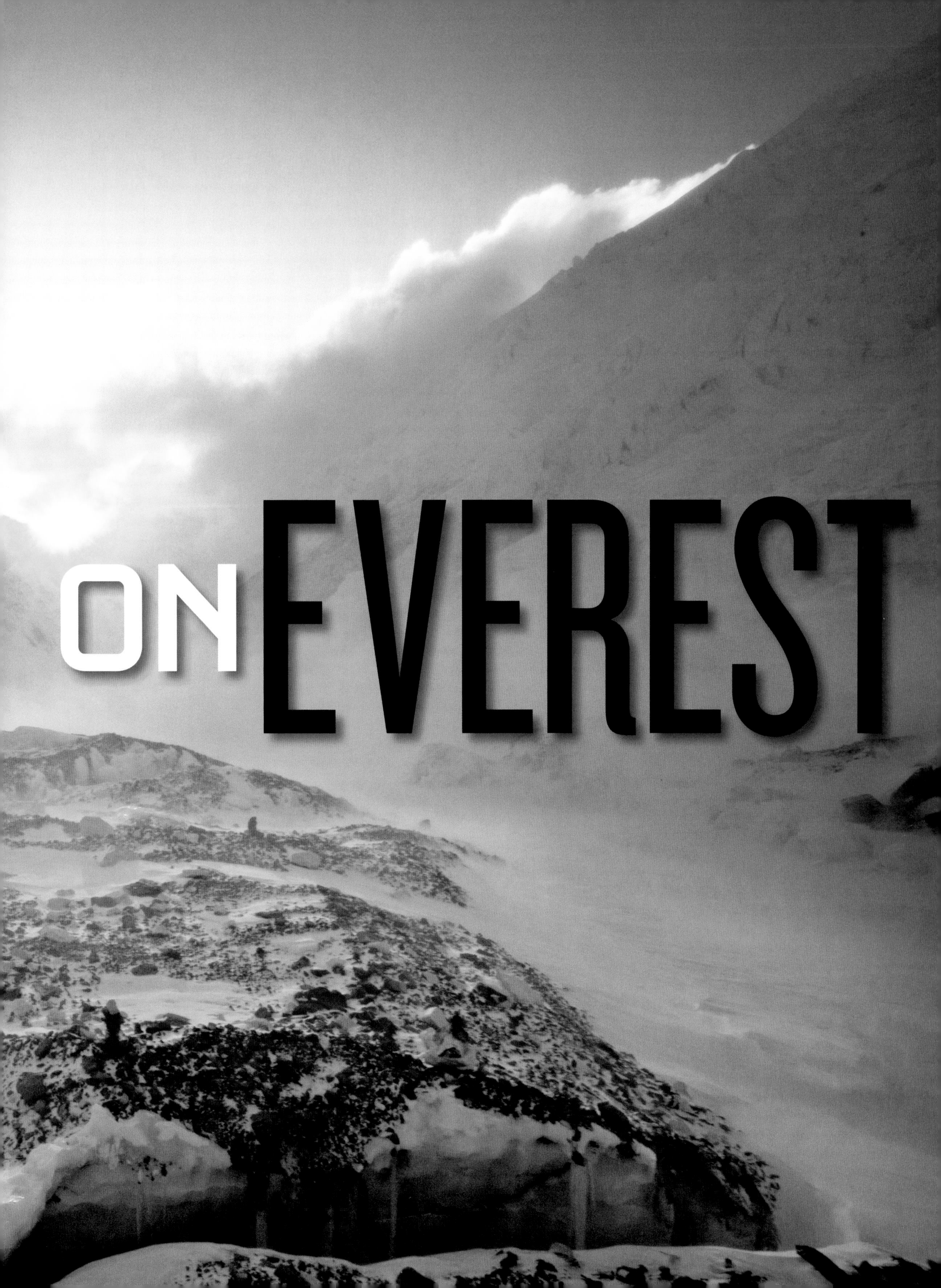

# BIG JIM WHITTAKER, 83 BUT UNSTOOPED, STILL BELIEVES IN THE MAGIC OF MOUNT EVEREST. "THE DAY AFTER I STOOD ON THE SUMMIT WE GOT A TELEGRAM FROM PRESIDENT KENNEDY."

Big Jim Whittaker, 83 but unstooped, still believes in the magic of Mount Everest.

"The day after I stood on the summit, we got a telegram from President Kennedy," says Whittaker, then recites it by heart: "For pushing human endurance and experience to their farthest frontiers . . . I know all Americans will join me in saluting our gallant countrymen."

On May 1, 1963, Big Jim, accompanied by Nepali Sherpa Nawang Gombu, became the first American to summit Everest. They climbed the South Col route, pioneered by New Zealander Edmund Hillary and Sherpa Tenzing Norgay on the first ascent of Everest ten years earlier. Three weeks later, in an unprecedented act of boldness, Whittaker's fellow teammates Tom Hornbein and Willi Unsoeld clawed their way up a completely unknown route, the West Ridge. On that same day Barry Bishop and Lute Jerstad made the second American ascent of the South Col. The two teams managed to meet up, but it was late in the day and they were forced to bivouac at 28,000 feet. Without tents, sleeping bags, stoves, oxygen, water, or food, no one believed they would survive.

**CONRAD ANKER ORGANIZES** bags at Base Camp, where climbers spend time getting acclimated before making a push for the summit. As the group's leader, he was faced not only with the logistics but also with making tough calls, such as when to try for the summit and when to wait.

"God, they were lucky," says Whittaker. "If there had been any wind, they would have all perished. It would have been horrible!"

All four men survived—although Unsoeld and Bishop would lose 19 toes and two fingers between them—and the '63 expedition became a tale of heroic American success.

# EVEREST 50 YEARS BACK

Fifty years later, Jim Whittaker, a mountain of a man, is walking to the world's highest mountain to commemorate the expedition's accomplishments. We meet on the trail in the wind and later sit down together in Pheriche, a dusty tourist village along the trek to Base Camp.

"Even though Hillary had climbed the South Col, there was nothing there," reminisces Whittaker. "We had to climb with our eyes wide open. We knew we had to get through the icefall. I hated it. That's where Jake was killed. But we had to put a route through it and we did. Up higher, I worked on the Lhotse Face for three weeks to get the route in. We led the route and then fixed our own ropes." When I respond that Everest is not what it was then, that today Sherpas fix lines from Base Camp to the summit and that 95 percent of climbers are clients, he nods, his good cheer undiminished.

"I still admire the people who reach the summit of Everest. It's hard work and there are still hazards," says Whittaker. "In a way, I think they're getting out of it what we got out of it back then. They're learning about the limits of their bodies. Nature is teaching them. They're learning about the beauty of our planet, and maybe this will get the bastards to protect it."

## TROPHY CLIMB

Commercial guiding on Everest, which began in earnest in the early 1990s, is little different from climbing assistance in the Alps over a hundred years ago. Ninety-nine percent of all ascents of the Matterhorn have been guided climbs. (Today most ascents of well-known mountains—Mont Blanc, Rainier, Denali, Kilimanjaro—are guided.) Namche Bazar has become the Zermatt of the Himalaya, and the Sherpas, like the Swiss before them, have profited handsomely from the influx of tourists, trekkers, and climbers. So too has the Nepali government, which charges every Everest climber a $10,000 peak fee. Surprisingly, the precipitous increase in client climbers on Everest has caused an equally precipitous decline in the statistical dangers of ascending the mountain. Back in 1986, when

> ## "KHUMBU HAD BECOME A TOURIST MECCA. ONLY THE MOUNTAINS HADN'T CHANGED."
>
> —CHARLES HOUSTON, AMERICAN MOUNTAINEER AND M.D. SPECIALIZING IN HIGH-ALTITUDE PHYSIOLOGY

Everest was a part of me before I was born. My father, Barry Bishop, first went to Nepal in 1961 to join the Silver Hut expedition on Ama Dablam, led by Sir Edmund Hillary. This was five years before my birth. The team wintered at 18,000 feet on the Mingbo Glacier collecting what would be seminal research on high-altitude physiology. My mother, Lila, led her first trek from Kathmandu to Everest Base Camp in support of the 1963 American Everest expedition and has been leading treks to the high mountains ever since. Mount Everest became an unknowing cornerstone in so many ways for our family: The mountain would define Barry's climbing career with his ascent in 1963, lifelong friendships were formed around the mountain, and countless adventures were launched with people Everest brought together. Our family would live in the mountains of Nepal in a tent for two years when I was a boy, and we have now worked, trekked, and climbed in the Himalaya for over 50 years. I simply can't remember a time when the Himalaya were not woven into the fabric of our family's life in one sense or another.

My father was a member of the American Mount Everest Expedition (AMEE) team, and I have been lucky enough to follow in his footsteps, reaching the summit of Everest in both 1994 and 2002 and attempting to summit the mountain via the West Ridge in the spring of 2012. This May marks the 50th anniversary of the first American ascent of Mount Everest, which was achieved by the AMEE. My own three expeditions to the mountain over the course of nearly two decades have given me a unique vantage point from which to reflect on the significance of the 1963 American team.

Porters carry expedition supplies on the 1963 AMEE

## GROWING UP EVEREST

As much as I picked Everest, the mountain picked me. My father was a photographer and scientist for the National Geographic Society and a renowned climber—and he was my hero. Growing up in his household left an indelible imprint on me. I remember watching *Americans on Everest* as a child, the first National Geographic television show ever produced. It documented that first American ascent of the mountain and was narrated by Orson Welles. I was captivated by the grand adventure of scaling the world's highest peak. This trip epitomized what exploration means to a young boy.

The men of the expedition were giants to me, larger-than-life climbers, and it was my great privilege to know them firsthand. I vividly recall scenes from our kitchen where my father and his cohorts would be telling stories about climbing and travel to far-off places. Unbeknownst to me, these men were all elite climbers, explorers, and scientists; I simply viewed them as my father's friends. Such moments filled a young boy with wonder and amazement for what awaited him in the world. As early as I can remember, following in my father's footsteps and climbing Everest was a dream of mine, and this quest seemed normal and simply part of the legacy that I was lucky enough to be born into.

Growing up with a father with a larger-than-life persona whose curriculum vitae fills page after page can be a tricky business. Pride and admiration for my dad were balanced by trepidation that his boots were so large it seemed a daunting task to fill them. I remember early climbing trips to the Tetons with my father when fear gripped me while on an exposed precipice or ridgeline. As I grew older, the fear of exposure gave way to a thirst for adventure and the unknown. Looking back, the greatest gift bestowed on me by my father was not technical instruction but an attitude that any adventure was possible and within your grasp as long as you took the first step and had commitment to the process. My first expedition to the Himalaya had lofty goals: to climb Mount Everest. And when asked about what I learned most from my father about climbing, the answer is not technique or skill but the unwavering belief in myself that I would be successful. Failure and triumph have marked subsequent expeditions, but it was the attitude my father bestowed on me from an early age that has been his greatest gift in the mountains.

And while these climbers were already heroes to me as a boy, it was not until I embarked on my own Everest climb at age 27 that I came to fully understand the significance of these men, and in particular their 1963 Everest

**Portrait of AMEE photographer Barry Bishop**

expedition. Climbing through terrain that I knew from pictures embedded in my memory from my youth, I was flooded with admiration and emotion. These early climbers hadn't only accumulated knowledge of the mountain after hundreds of ascents over decades of climbing. They earned their knowledge of the route one foot at a time. I remember struggling with the weight of my own oxygen equipment as I labored up the Geneva Spur to Camp IV at 26,000 feet. There on the South Col I picked up a discarded bottle from the AMEE's era of rudimentary equipment. It weighed three times as much as my modern apparatus. Even with the psychological barriers of the unknown removed and the benefit of more than 30 years of technological advancement, the climb pushed me beyond what I thought was possible both psychologically and physically.

## CONNECTING

The first time I approached Everest's summit in '94, tears ran down my face as a deep sense of connection to my father welled up inside of me—a connection as a climber I had finally earned a right to call my own. I now had a true glimpse of how strong and bold those climbers were in 1963, a time when there was no simple formula for success and every step on the mountain was gained through sheer will and fortitude. Their lives hung on every decision they made, with no preordained template to guide them to the summit. Tragically, my father was killed in a car accident only four months later, robbing me of the opportunity to share and reflect with him as the first American legacy to reach the summit.

Nearly a decade later, I returned to the mountain with the opportunity to partake in a film for the National Geographic Society, *Everest: 50 Years on the Mountain.* The film celebrates the 50th anniversary of Sir Edmund Hillary and Tenzing Norgay's first ascent and is told through the eyes of the sons of these pioneer climbers. When asked to join the project, I jumped at the invitation to partake in a film that would continue the legacy of my father and his peers. And just as the original film influenced me as a young boy, partaking in a film that might influence my own two sons made me proud to be part of the American Everest legacy.

AMEE climbers pause at 24,000 feet.

Our climbing goal was to repeat Willi Unsoeld and Tom Hornbein's epic first ascent of the West Ridge, but poor weather and high winds hastened our retreat from the West Shoulder. We were able to redirect our efforts and summit via the South Col route, and once again, I literally and figuratively followed in my father's footsteps to the summit of the mountain, carrying both the American and National Geographic flags that my father had taken with him nearly 40 years earlier. The memory of that summit day is bittersweet for me; I remember Peter Hillary making an emotional call to his father from the summit, and while elated from standing on the top of the world, Peter's connection with his father made me acutely aware of my own father's absence at such a significant moment for me.

The allure of Everest is quite strong, and after failing on the West Ridge in '02 I felt I still had unfinished business on the route. The aesthetics, commitment, and climbing style embodied by the line appeal to me on every level as a climber, so I was glad for the opportunity, in 2012, to join a 50th anniversary commemorative climb of the West Ridge. This team was quite experienced, with 11 summits of the mountain between the four principal climbers. But once

again the conditions on the route were unfavorable and forced our retreat after relatively little progress above Camp II. Each day that our veteran team labored to gain ground on the route and found itself bogged down, I kept wondering with amazement how the original American climbers were able to push the route successfully into the unknown so effectively. Nearly 50 years after first ascent of the West Ridge, the route's significance has withstood the test of time. The technical nature of the climb and the commitment it demanded were unparalleled in 1963, and it still remains a coveted achievement in the Himalaya, as evidenced by the mere handful of successful ascents in the 50 years since.

## THE MEANING OF AMEE

Not only has the first American ascent withstood the test of time, but within the context of the era, the AMEE is even more significant: 1963 was the age of Camelot, when Kennedy championed the merits of the "vigorous life." At this point, America had lost the first step in the space race to the Soviet Union's successful launch of the first man into space. A wave of patriotism swept through America, and it was clear that no one—not mountaineers, politicians, nor the general public—wanted to see America lose another race to the Soviets. The AMEE team was assembled, and the national support it received was not only unparalleled at the time but has not been witnessed since.

The budget for the AMEE expedition was $405,000. This was a huge sum of money for its time and would be equivalent to over three million dollars today. To add perspective, our expedition budget in '94 was $80,000, and this included a mandatory $50,000 peak fee for the ministry of tourism. But even more impressive than the monies raised were the breadth and volume of support that came from myriad benefactors. It is hard to comprehend in this day and age—when expeditions receive little public attention or support—the wellspring of backing the AMEE received. Private individuals donated personal funds, companies and stores donated tons of equipment, and a host of government organizations backed the endeavor. NASA, the Department of Atomic Research, the Navy, and the Air Force all donated substantial funds to the expedition in return for scientific research. National Geographic was also a major supporter in return for the photographic and film rights to the expedition. So captivated was the nation with the expedition that President Kennedy received the team at the White House and *Life* magazine ran a cover story featuring the climb.

The objectives of the AMEE were lofty, to say the least. The primary climbing goal of the expedition was to place a man on the summit via the Southeast Ridge. This was accomplished on May 1, 1963, when Jim Whittaker and Sherpa Nawang Gombu stood on the top of the world. This feat alone marked success for the team, but further goals were in store. The idea of climbing the never-before-attempted West Ridge, traversing the mountain, and

descending the Southeast Ridge became the secondary objective. The West Ridge is a route that truly embodies the spirit of mountaineering—it is visionary, technically difficult, and requires a tremendous amount of commitment. In 1963, the mere idea of ascending via the West Ridge was at least 30 years ahead of its time. Willi Unsoeld and Tom Hornbein's success on the route was to mark America's greatest Himalayan achievement to date.

Not only did Hornbein and Unsoeld climb the West Ridge on May 22: They met my father and Lute Jerstad near the summit of the mountain and then began to descend through the night. Nightfall caught them out in the open and they were forced to bivy at over 28,000 feet without stove, tent, or oxygen. The team huddled together exposed on the Southeast Ridge with no other option except to wait for the light of morning. Nobody had spent the night out at this altitude before, and it was uncertain whether survival

**AMEE climbers rest high on the mountain, their '60s-era equipment in full view.**

was even possible. Twice I climbed past the site where the foursome spent the night unprotected below the summit and on both occasions I remember being overwhelmed with the notion that this was the location where my father struggled to stay alive through the night. I often wonder what occupied my father's thoughts as he waited for the cold and blackness to give way to the morning light.

## LIVING SCIENCE

The other objective of the AMEE was to perform scientific research while on the mountain. In essence, the AMEE was its own scientific lab. Significant research and experiments were performed in the disciplines of glaciology, solar radiation, physiology, psychology, and sociology. Every member of the team contributed in some form to the research component of the expedition. In 1963 there was not a great deal of data on high-altitude physiology. The AMEE not only provided the perfect lab but all of the specimens to study as well. The results from the AMEE provided concrete documentation of the effects of altitude, particularly above 8,000 meters, and this body of knowledge has served as the foundation for all subsequent altitude research done in the United States.

The Navy and Air Force were quite interested in the social and psychological aspects of a team subjected to prolonged high stress and extremely cold weather. Their interest was prompted by the U.S. Defense Department's belief that the country would most likely be involved in a conflict where soldiers would be stationed in very cold, stressful, and remote conditions. Interesting as they were,

**John F. Kennedy presents Hubbard Medal to Norman G. Dyhrenfurth.**

the results of the research were most likely biased, given the exceptional bravery and strength of character of the expedition members.

Assembled by expedition leader Norman Dyhrenfurth, the members themselves were the best and most experienced mountaineers in North America in their time. The list of members reads like a who's who of the '60s era. Just thumb through any guidebook from almost any climbing destination in the United States and the names Corbet, Pownall, Emerson, Unsoeld, Jerstad, et al. can be found in connection with significant ascents all over North America. Not only were the AMEE members exceptional climbers, they were highly educated and professional. The team had three M.D.'s, five Ph.D.'s, and five M.A./M.S.'s. (Three of the members were working on their Ph.D.'s on the expedition.) These men were from the renaissance mold of climbers; their avocation of climbing was separate from their vocations, which only makes their feats on Everest even more impressive.

Fifty years ago, this expedition was undertaken with a mixture of excitement, adventure, and optimism. When we look backward we have the benefit of seeing the risk of exploration and the light of its results. The tangible results of the AMEE are quite clear—six members on the summit, a new route on the mountain, and a massive body of scientific research produced. Less tangible is the tremendous influence these climbers and their expedition had on subsequent generations to follow. Virtually every climber who ventured forth into the mountains in the generations following the 1963 climb read either Hornbein's *Everest: The West Ridge* or Ullman's official account of the climb, *Americans on Everest,* if not both. These books influenced countless climbers on many levels.

## SETTING THE BAR HIGH

Most important, this team set the benchmark of what was possible for American climbers in the Himalaya. After

At 25,000 feet, a push toward the summit in 1963

reading these accounts, one comes away with the belief that anything can be accomplished; and one comes away inspired to accept the mantle of these mountaineers' legacy and follow in their footsteps. The world of the known expands only when we place our faith and support in those brave enough to take giant steps. We would be nowhere, essentially, without the bravery of our explorers, and that's why we are celebrating this anniversary. In essence, any American climbing in the Himalaya is climbing in the footsteps of giants. And among these giants are all the members of the 1963 American Mount Everest Expedition.

A lifetime of climbing and 46 years of age brings greater perspective in the mountains—at least it has for me. No longer armed with the invincibility and strength of youth, my last climb opened my eyes even more to the AMEE's achievement. A chorten to commemorate my father sits behind Tengboche monastery alongside memorials for his friends and fellow climbers Gil Roberts, Lute Jerstad, and Jake Breitenbach. It's a powerful sight with a beautiful view up the valley and Everest looming above all else. I visit these chortens whenever I'm in the Khumbu to string up Tibetan prayer flags and to have a drink with "the boys." Tears are shed, and once again I feel like the kid in his kitchen surrounded by his heroes, listening with amazement and wonder to tales of Everest.

—**BRENT BISHOP** **Brent Bishop is a climber and environmentalist and son of Barry Bishop, member of the 1963 American Everest expedition. In 1994 Brent Bishop founded the Sagarmatha Environmental Expedition, which has since helped to remove thousands of pounds of trash from Everest.**

THE 2012 LEGACY Climb team at the beginning of the expedition (above)

CONRAD ANKER (BELOW, at left) talks with famed Italian alpinist Simone Moro. Moro pulled out of his 2012 attempt to summit without supplemental oxygen, citing concerns about the crowds.

I first attempted Everest with a team of accomplished alpinists (we reached 8,000 meters on the North Face before being turned back by avalanches), if you were a summiter, you had about a 10 percent chance of dying—worse odds than going to war. Now it's under 2 percent.

Whittaker, a former guide himself, concludes our conversation with a positive perspective toward the commercialization of Everest: "It's just fine if people are paying $10,000 to climb Everest. More power to them. They could be killing endangered species in Africa, going for some disgusting Boone and Crockett trophy."

Trophy. The word will be lodged in me like a sliver for the rest of the expedition.

## THE LEGACY CLIMB TEAM

Wiry Hilaree O'Neill, redoubtable Dawa Yangzum, and I, aging but unrelenting, arrive at Everest Base Camp on April 10, joining the rest of our team. Like Jim, we've all come to the Khumbu to celebrate the 50th anniversary of the first American ascent of Everest. Led by Conrad Anker, renowned alpinist and Everest veteran, the 2012 Everest Educational Expedition is composed of nine climbers, eight high-altitude Sherpas, one high-altitude Sherpani, three communications technicians, and seven Base Camp staff. Our Sherpas—Ankagi, Danuru, Jangbu, Lakpa Bhote, Mingma, Panuru, Sonam Dorje, Tendi, and young Sherpani Dawa Yangzum—are some of the most experienced in the business. Our sirdar (Sherpa leader), Panuru, has summited Everest 9 times; Danuru, by far the strongest member of our team, has summited 13 times.

The Western climbing team is composed primarily of The North Face professional athletes. Besides Conrad, 49, there is Hilaree O'Neill, 39, ski mountaineer; Kris Erickson, 38, photographer and ski mountaineer; Cory Richards, 30, mountaineer and photographer; Sam Elias, 29, elite mixed ice climber; and Emily Harrington, 25, a competition sport climber. Rounding out the team are National Outdoor Leadership School (NOLS) field manager Phil Henderson, 48; Montana State University geologist Dave Lageson, 61; and myself. Combining our experience, we've totaled over 200 expeditions, but only Conrad has previously summited Everest.

Conrad and Cory will attempt the West Ridge, the rest of us the South Col. Not surprisingly, merely climbing the mountain is insufficient for Conrad. Hence,

# "NOW WE HAD INDEED LEFT THE WORLD BEHIND."

–JOSEPH F. ROCK, *NATIONAL GEOGRAPHIC*, 1930

**ALL APPEARS TO** be quiet in this night shot from Base Camp.

as in the '63 American expedition, we are also conducting geological and physiological research, as well as creating an elementary school science curriculum.

The team, almost 30 members in all, comes together for the first time at Base Camp. Spread across the rock-strewn, ever shifting Khumbu Glacier, the 17,600-foot Everest Base Camp is probably the most famous, albeit ephemeral, town of its size in the world. The pre-monsoon population is perhaps 900 temporary inhabitants: some 300 clients and guides, a matching number of Sherpas, and another 300 Base Camp staff and porters, not to mention the constant flow of yak trains, their bells ever tinkling in the frigid air.

## BASE CAMP CITY

Just like every town, Everest Base Camp has neighborhoods. Our camp lies in the southern burbs, near the huge Himalayan Experience outfitters camp. Himex, as it is commonly called, boasts the highest, tackiest nightclub on earth—the Tiger

Dome—a massive, hard-skeleton geodesic bubble with a 30-foot tiger rug spread over a flexible dance floor, big speakers, fake tiger fur draping lounge chairs, and a free-flowing bar. We will burn a few brain cells in the Tiger Dome—Conrad and Emily swinging like monkeys from the ceiling, Hilaree and friends diving in tandem off the bar into the crowd—before crawling back to our tents in the wee hours.

On the main street going into town is the International Mountain Guides borough to the left, and Mountain Experience to the right. Farther along is the ever busy helipad, a circular patch of rocks, and the large white tent of Everest ER, the medical center, founded ten years ago by Dr. Luann Freer. Downtown can be found Mountain Guides, the Korean University, Happy Feet, the Indian Military and two dozen other expeditions of unknown provenance. Each hood looks fairly similar, with a long, rectangular mess tent surrounded by 10 or 20 orange or yellow dome tents and a couple vertical blue or green outhouses. Prayers flags flap like community banners from the stone chortens of each neighborhood.

Despite the bad press over the years, Everest Base Camp is a fairly tidy place. The SPCC—Sagarmatha Pollution Control Committee—is the Base Camp's city council, and they make the rules. Tent outhouses are fitted with plastic barrels and all human waste is transported down to the village of Gorak Shep. Unlike at higher camps, here at Base Camp every team is required to pack out what they pack in.

Daily life in BC City is surprisingly civil and leisurely. You can get the Internet, iPhone reception, and the occasional cook whips up a birthday cake. As at summer camp, the staff provides meals so the climbers can rest, recover, listen to music, take pictures, send e-mails, write blogs, prattle on the phone, or just stare, mouths agape, at the avalanches roaring off the surrounding flanks of Pumori and Nuptse. We even have afternoon tea, a classic sahib tradition.

The majority of the inhabitants here are not hard-core climbers. They're clients with the dream of climbing Everest. Many of them have been successful in another walk of life—business, medicine, technology—and, knowing full well how long it took them to gain their own skills, are happily paying for the expertise of a guide. It's easy to mock these aspirants, but there will always be people with the cash to climb Everest, and as Whittaker noted, they could be blowing their bucks on more dubious pursuits. Instead, these neophytes are willing to risk their lives passing through the jaws of Khumbu Icefall, suffer the smashing headaches and projectile vomiting and weeks of intense fatigue, just to stand on top of the world. That doesn't make them mountaineers, but it does speak to their conviction.

The first few days at Base Camp are spent acclimating, and many Westerners either get the "Khumbu crud" or spend a lot of energy doing all they can

SUNIEL, PART OF the Sherpa support crew at Base Camp, shreds cheese for a meal. Climbers find that eating enough calories can be difficult when altitude has numbed the appetite and the human body needs to burn more calories to stay warm.

**LAMA NIMA RITA** leads a puja (below) with the team in Base Camp, a Buddhist tradition as an Everest expedition begins. During the ceremony, Sherpas and climbers pay homage to the mountain deity and cover an altar with offerings. The lama blesses objects such as crampons, and prayers are sent off to the gods.

**LIKE A SURFER** climbing a wave, team member Sam Elias (opposite, at top) practices a night climb of an ice pinnacle near Base Camp as a way to keep his climbing skills honed. While this climb is not part of the route to Everest, exercises like this are important to retain overall physical and mental fitness.

to avoid it. The sometimes squalid teahouses along the week-long hike to Base Camp can be petri dishes for disease, and few escape some form of gastritis. If you don't get the crud, you're almost certain to get the "Khumbu cough," an irritation of the bronchial tubes caused by the extremely dry air. A hacking cough is the ubiquitous symptom, and soon everyone is wearing buffs over their faces like bank robbers.

# FACING THE GLACIER

The first obstacle on the south side of Everest is the infamous icefall. The Khumbu Glacier, like a gargantuan bulldozer, plows down off the Lhotse Face between Everest and Nuptse, pushing walls of blue ice before it. Dropping over a cliff just above Base Camp, this mile-wide glacier splits apart, shattering into building-size blocks and seracs as tall as church spires.

Every year the route through the icefall is set by the "ice doctors," a small hit squad of Sherpas who take mortal risks to navigate the safest passage, putting up ropes in the steep sections and stretching ladders across the abyss-like crevasses. Nothing more than a massive, moving river of ice, the Khumbu Icefall is constantly

Mount Everest isn't supposed to be a difficult mountain to climb. "Climbing" is often not even used to describe the act of reaching the summit of the world's tallest peak. "It's just walking. It's not hard"—that was something I'd often heard through the years. Growing up in the climbing community gave me an opinionated perspective toward a place I'd never even visited or bothered to learn about. When non-climbers asked me if I ever dreamed of climbing Everest, I would snidely reply "Um, no. That's not the kind of climbing I do."

Well, I'm now eating my own arrogant words. I went to Everest this year on a trip with The North Face. Conrad Anker asked me to be a part of the team, handed me the opportunity of a lifetime, and I took it. I had little knowledge of the alpine climbing world and zero experience in high-altitude mountaineering. I had a very skewed view of the uniqueness of Everest itself and the intense polarizing effect it has on the community I was a part of.

### THE STRUGGLE

Everest is a controversial place, full of both real-life danger and ego-crushing criticism. Climbing Everest was the most personal struggle I've ever had to undergo during a consecutive period of time. I've never been as sick as I was with a respiratory infection when I first arrived at Base Camp. I've never fought so much physically to keep pushing, taking steps, and enduring the exhaustion, extreme heat, and bitter cold. I walked by dead bodies, human souls who'd lived just four days prior, and left this world in pursuit of the same goal I was trying to achieve. I was afraid a lot of the time. Never before have I faced a truer reality: that my own life could be taken away from me by circumstances out of my control, together with the unsettling knowledge that it was my choice to be there, but for what? I fought intensely with my own mind on a daily basis to justify this mission to myself despite the danger, death, and even the harsh criticism I was receiving for even setting foot on the mountain in the first place with no previous high-altitude mountain experience.

There are some very glaring negative environmental and social impacts of climbing Everest, especially considering the commercialization of it all. People who have never even set eyes on the Khumbu Valley know these things. But there are other aspects that are often overlooked or ignored, overshadowed by the negative ones. My own personal journey on Everest was full of negativity and struggle, but also triumph and success; and I arrived home a different person than when I left. I saw another side to the place that so many criticize and condemn, and what I remember most is the beauty and passion that exists there in so many forms.

### THE NOISES

I remember listening to avalanches all day and night from my tent at Base Camp, watching their uncontrollable power and violent strength and feeling simultaneously afraid of and fascinated by the beauty of the noise and energy they emit. Then there was my first trip through the icefall at 3 a.m., and the acute fear that made my stomach plummet and brought tears to my eyes at the very real danger of it all. I remember the intense heat of the Western Cwm, watching the snow melt and evaporate into steam in a matter of seconds after I put it down my shirt in an effort to cool my boiling blood. I remember listening to the wind barrel down off of the summit of Everest from Camp II, like a freight train with no brakes. It would reach my tent moments later and suddenly I'm in the middle of a hurricane, the nylon ripping and floor trying to lift my helpless body off the ground. I remember the first night we slept at Camp III, the unusually calm evening that welcomed us after so many torturously windy ones. The glorious sunset I watched that night made me feel like the luckiest person on Earth. I remember summiting early in the morning on May 25 in the nuking wind and -50° temperatures and trying to consciously be aware and appreciative of where I was and what I had done—but also realizing that reaching the summit had the smallest fraction of significance to me in the grand scheme of what I had seen and felt and how I had changed as a person on this journey.

### THE PEOPLE

I remember the mind-blowing strength and kindness of the Sherpa people, and how they always managed to be in good humor despite the massive amount of effort their jobs required. I remember the relentless respect and love I felt for both our Sherpa team and for my Western teammates. They all became my family, individuals whom I will never forget and with whom I will always be bonded.

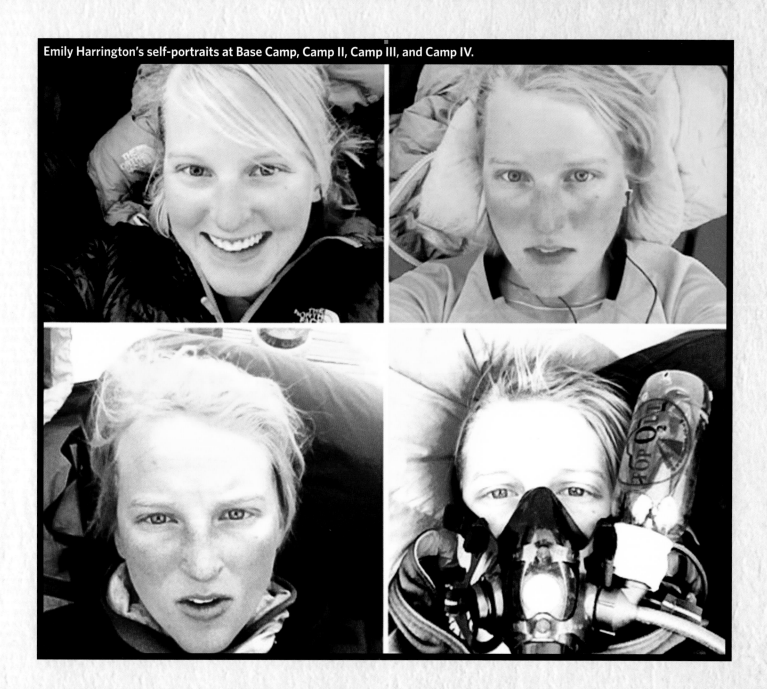

Emily Harrington's self-portraits at Base Camp, Camp II, Camp III, and Camp IV.

I also remember the bizarre community at Base Camp, like a small town with all the same drama and gossip, but also full of real people, with jobs and families and purpose. Passionate people with positive intentions who were there simply to fulfill a dream and to experience the enormity of the mountains. Like them, I too felt the allure of being in a place so much more vast and powerful than we can comprehend. It is overwhelmingly humbling and puts us in our respective places as human beings. Maybe that's why people go there in droves. In addition to the tangible goal of reaching the top of the world, they're searching to experience a place that's so much greater than themselves, to struggle and suffer and fight and discover who they are, much like I did.

I remember all of those things, I am immensely grateful for them, and I will never forget them.

—EMILY HARRINGTON  A North Face athlete on the Legacy Climb team, Emily Harrington is currently the only American, male or female, consistently placing in the top five in the Climbing World Cup competition circuit.

shifting, groaning, and collapsing, and if you happen to be there at the wrong time, you will be swallowed in a crevasse or crushed by a falling serac. Three Sherpas died in a single accident in the Khumbu several years ago. This spring a line of prayer flags demarcates the beginning and end of the most perilous section.

According to actuarial research, more deaths on Everest occur coming off the summit than wending through the icefall. Nonetheless, passing through the maw of the Khumbu is a morbid game of Russian roulette, so we try to do it in the dark, before the sun warms the ice and it begins calving. We leave Base Camp at 2 a.m., following the beams of our headlamps on our helmets like miners. On my first trip I'm paired with Danuru. When we reach the first line of prayer flags, the starting gate to the death zone, he allows me to catch my breath, then says, "Now we go!" and takes off running uphill, literally, carrying a 60-pound pack at 18,000 feet, his crampons scraping on the boulders of blue ice.

What does it feel like to go through the Khumbu Icefall? Imagine walking on railroad tracks through a dark mountain. You know a train comes roaring down the tracks at random times. There is no way for you to get off the tracks if a train comes. Sometimes the tracks go over a rickety bridge, with a bottomless pit below, and you have no idea if the bridge will collapse when you walk out on it. But there's nothing you can do. So you just keep walking, hoping you'll be lucky, hoping the train doesn't come until you've passed through the mountain.

Carrying food and fuel, tents and oxygen bottles, our Sherpas will hike through the Khumbu 30 or 40 times in one season, whereas we Westerners hope to only run the gauntlet perhaps eight times. The days we are resting in Base Camp, the Sherpas are humping loads through the Khumbu for us. It's a mercenary business that raises real moral questions.

## ACCLIMATING

Through the month of April, Camp I, at 19,700 feet, and Camp II, at 21,200 feet, are established. Camp I, a windy, gorgeous place, lies on a bone-white rib of glacier between deep, wide crevasses. The Sherpas do not stay at Camp I, so it's more like a climbers' camp in Alaska or Peru—you do your own cooking and look after yourself. Camp II, as

SURROUNDED BY ICE and snow (previous pages), the team makes its descent through the Khumbu Icefall and back to Base Camp. Climbers travel from Base Camp up to Camp I and Camp II, spending a few days at the higher elevations as a way to acclimate.

HILAREE O'NEILL (ABOVE) climbs the base of Lhotse Face. Everest's dry conditions during the 2012 season meant that stretches of the climb were on slick, hard ice instead of the more typical snow cover. O'Neill is heading to Camp III during an acclimation trip.

ugly as a city dump, squats on the left lateral moraine, tents pitched among the hundreds of shit piles left over from previous expeditions. We tramp up to Camp I for several days, then back down to Base Camp for a few days, then back up to Camp II, then back down to Base Camp. This monotonous yo-yoing is essential for acclimation.

The boredom of days at Base Camp is relieved by the arrival of Dr. Bruce Johnson and four researchers from the Mayo Clinic's Human Integrative and Environmental Physiology Lab. They've come to study the connection between high-altitude acclimation and one of the most common diseases in America, heart failure.

"We have data that show many heart-failure patients get constrictions in the lungs," explains Johnson, "which could be similar to what happens to climbers at altitude."

Heart-failure patients often have pulmonary congestion: Their lungs easily fill with fluid. Well-acclimated climbers have lungs that successfully balance pulmonary fluids. Johnson and team want to understand the specific mechanisms that facilitate this process.

"Another similarity between high-altitude climbers and heart-failure patients," says Johnson, 54, "is that they both chronically hyperventilate. This may cause the brain to inhibit breathing, and in some cases trigger sleep apnea." Why and how this happens is another subject of scrutiny.

## GLACIAL GUINEA PIGS

To gather heart- and lung-function data, we all have a tiny computer, smaller than the smallest cell phone, taped to our chests, which will remain in place the rest of the expedition. This multi-thousand-dollar device is wired to two electrodes attached just below our nipples. A high-resolution accelerometer and miniature electrocardiogram, the machine that measures heart rate and the interbeat interval takes 400 readings per second. Heart rate variability is one measure of stress on the system.

Over the course of a week, the Mayo team performs a series of tests on us and themselves. Blood is taken, fat pinched with calipers, and a jelly-smeared ultrasound wand slid back and forth across our chests. Machines measure our lung size and airway function. Our cognitive skills (or degeneration thereof) are measured with computer games similar to those once given to astronauts. In

## THEN & NOW INSIDE THE TENT

| 1953 | 2012 |
|---|---|

Which tent scene suits your style? The 1953 pig pile for human warmth, or the NASA-monitoring intensity of today? The 1953 team is actually listening to the coronation of Queen Elizabeth—radio was available—while the communications tent in the 2012 expedition includes Wi-Fi with its full array of amenities.

one, we stab a plastic pen on the screen whenever a square appears—this is to measure hand-eye coordination. In another we match the color of a word—for instance, the word *red* printed in the color green—with the name of the color—not the word—below. Finally we do a grip-strength test and a step test, jumping up and down on a small stool like Jane Fonda.

After both torturing and entertaining us, Johnson and his Mayo cabal pack up and head home to Minnesota to crunch the numbers. (They take our geologist, Dave Lageson, with him. He is not only sick but has to attend to family issues. Rocks collected on the summit by our Sherpas will be sent to him for analysis.)

"Our goal is to learn as much as we can about the physiology of the cardiopulmonary system under extreme stress," summarizes Johnson as he departs, "and then apply these findings to treating heart-failure patients."

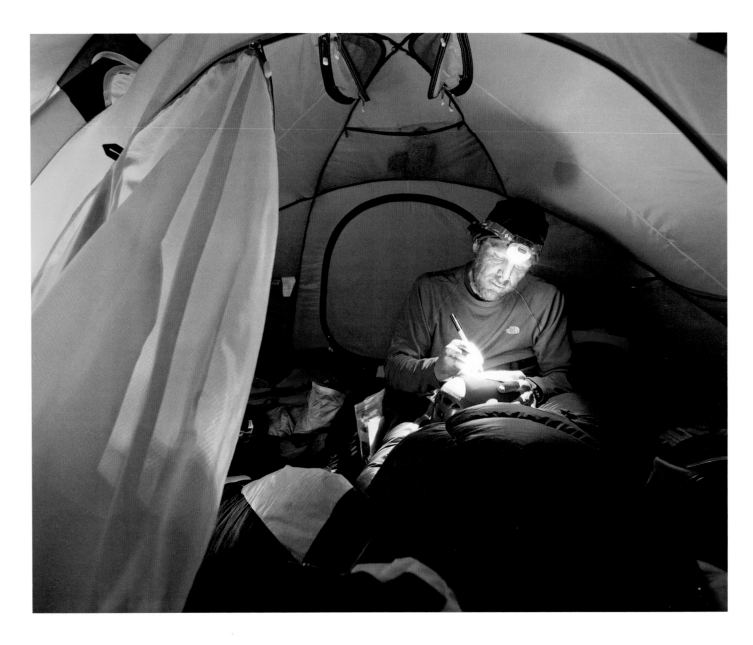

Sam Elias rests during his descent from the summit.

Thirteen thousand feet above sea level. That is about the highest elevation that I had ever been to before going on this Everest expedition. Though I am a climber and a lover of the mountains, mountaineering has never been a realm of the sport of climbing that has entered my mind. I dream of steep rock and ice, and mostly as single pitches. Having been offered an opportunity to attempt the world's highest peak as a part of the National Geographic/The North Face Legacy expedition team, I was compelled to expand my understanding of climbing in the big mountains and of being on a proper expedition. Still, I was vexed by all of the uncertainty that I knew I would experience in the commitment of an entire spring season. As I reflect back now, just a few months from standing on the summit of the world, I am certain that the experience did change my life.

## INNER PURPOSE

It was a long trip, and very challenging at every level of my existence—mentally, emotionally, spiritually, and of course physically. As it was my first major mountain expedition, I had no skills to understand what or how I should feel internally. The days were long, and in such a basic and raw and extreme environment, there is nowhere to escape. There are no distractions to remove the mind from where it is: quite simply, a place where people weren't meant to be. Our mortality and true fragility are so apparent there, so much so that our bodies are literally killing themselves. Unable to obtain enough calories, they start consuming themselves. The immune system is depleted, running on fumes, so sickness is inevitable.

I had two crippling bouts. One was a gastrointestinal nightmare, and the other was a throat and upper respiratory infection from hell. Both of which required antibiotics. However, there is always a silver lining.

This experience allowed me to sift through my life by forcing me to be present, live slowly, and confront reality. It exposed my true self over the weeks, pulling back layers like those of an onion. And it forced me right up against the tough questions: Who should I be? What is this life for? What do I want?

Perhaps it would have been better not to allow my mind to go so deep, especially given all of the responsibilities of our important expedition—but, as I said, I had no reference for what I should feel or think. I actually believe that part of the reason for my being there was to provide a fresh, unbiased perspective. I have no doubt that what happened to me there was because I needed it. Thankfully, we delivered on our obligations and were fortunate to summit as well, but in retrospect those things were like the cherries on top.

Most important, Everest allowed me to look at my life differently and to reconsider it. For this I will be eternally grateful both to those who believed in me enough to allow me to go . . . and to the mountain itself.

—SAM ELIAS  A dedicated alpine skier, sport climber, and winter climber, Sam Elias pursues challenges all over the world. He is a North Face athlete and was a member of the Legacy Climb expedition team.

# "I AM NOTHING MORE THAN A SINGLE NARROW GASPING LUNG."

-REINHOLD MESSNER, ITALIAN ALPINIST WHO, WITH PETER HABELER, MADE THE FIRST ASCENT OF EVEREST WITHOUT SUPPLEMENTAL OXYGEN, 1978

## OMENS OF THE SEASON

By the end of April, three deaths have occurred on the south side of Everest, all Sherpas. One died of a stroke, another of alcoholism, and one, failing to clip into the safety lines, fell into a crevasse in the icefall.

On April 26, half of our team is collected at Camp II and the other half at Camp III when a coxcomb of ice on the west ridge of Nuptse explodes, sending a monstrous avalanche down into the Western Cwm. A hundred feet deep, blasting along basketball-size boulders of ice, the avalanche tears right across the route from Camp I to Camp II. At certain times during the day, 50 to 100 people are traversing this path. Miraculously, only one person is caught in this avalanche. Blown into a tomblike crevasse, Nima Sherpa breaks three ribs and two vertebrae and bites through his tongue. Rescued by the Benegas brothers, Argentinian Everest guides who regularly save climbers, Nima is choppered out to Kathmandu.

### CORY'S COLLAPSE

The following morning, Cory and Conrad set off to recon a route up to the West Shoulder, the beginning of the West Ridge. They reach a high point of 23,000 feet, dodging falling rocks whistling by like missiles, before deciding that the route is

simply too dangerous to climb. Turning around, Cory begins to experience chest pain and shortness of breath. By the time they have dropped back to Camp II, Cory is gasping, dry heaving, and desperately weak. Over the next hour, we have Cory rest, relax, and drink. Visibly fatigued, his eyes anxious, Cory keeps repeating, "I just can't catch my breath. I feel like I'm suffocating."

We bring in two doctors from another team to examine Cory. Although his heart rate and arterial oxygen saturation are normal and he has no gurgling, wheezing, or signs of fluid in his lungs, his respiration rate and lower-left-side chest pain will not go away—the contradictory symptoms lead the docs to a diagnosis: "He may have a pulmonary embolism," says one. "A life-threatening emergency," says the other.

Eight Sherpas from Camp II, including Panuru and Danuru, volunteer to drag Cory down to Camp I. The Sherpas, Cory's photo assistant Andy Bardon, and I run down the Western Cwm, pulling Cory like a child in a toboggan. Black, low-slung clouds at Camp I eliminate the opportunity for a chopper evacuation, and carrying Cory through the icefall will take hours. Instead, cranking his oxygen up to four liters per minute, Cory, with the help of Panuru, stumbles down through the icefall and is immediately flown out of Base Camp for medical attention.

TEAM MEMBER AND photographer Cory Richards started to feel short of breath at 23,000 feet (7,000 meters) while fixing ropes with Conrad Anker. He descended to Camp II, where doctors gave him supplemental oxygen and then suggested a descent. Richards then made his way to Base Camp, sometimes on a Sked rescue stretcher (far right), but also under his own power through the treacherous icefall (opposite, at left). Doctors performed additional tests at Base Camp, from which he was evacuated to a medical facility near Lukla, Nepal.

TEAM MEMBERS DIG out a new platform for their tent at Camp III (below) after a serac avalanche filled in one of their tent sites with debris. With ten-story-tall seracs looming over them, the team pushed out from Camp III and on to the summit.

TEAM MEMBERS, WHO dubbed themselves the South Col Five, make their way up the Lhotse Face (opposite) using fixed ropes. They'll reach Camp III, halfway up the face, where they'll sleep in a stark but sublime landscape.

# THE TEAM PRESSES ON

This leaves Conrad without a partner for the West Ridge attempt. Flamboyant Italian climber Simone Moro, veteran of some 40 Himalayan expeditions, briefly considers teaming up with him. However, after analyzing aerial photographs of the West Ridge's notorious Hornbein Couloir, which is no longer snow but blue ice and exposed rock, he declares any attempt suicidal. "I didn't come here to die. On this route, this year, you will die." Both Conrad and Simone reluctantly decide to climb the South Col.

In the past half century, the physiognomy of Everest has changed dramatically. Back in the days before global warming, it was a snowy peak. Today the snowfields and glaciers have receded substantially, uncovering loose, friable rock. Due to an extremely dry, windy spring, during April both the Khumbu Icefall

Being on Mount Everest made me realize what a great life I have. It made me think of my own mortality and what is important.

For me, family and mountains are very important. It helped to be able to call home and talk with my four-year-old daughter to wish her a happy birthday and to hear the voices of her and my wife.

It made me realize what a small creature I am in the bigger picture of the planet. I would often notice the small differences of life at Base Camp, such as the glacial movement underfoot and the sound of serac fall in the night. To have the opportunity to see this small part of our planet close up was remarkable.

## FULFILLING GOALS

I quite often found myself hiking in the icefall alone and being completely comfortable. Being alone in the mountains is something I have always been comfortable with. However, being alone on the highest mountain in the world, where the risks are high, and still feeling content—that will always be a memorable part of the expedition. It told me that all the years of being in the mountains had in fact prepared me for this expedition.

I really tried to not worry about what happened. If I reached the summit, great: That would be the icing on the cake. My main focus was to make a positive impact and contribution to the expedition, the team, and the others around me. I feel I was able to accomplish all of my personal goals—with the exception of reaching the summit. There were so many other important things besides just getting to the top, though. I remember thinking about the three minutes of my life when I couldn't walk or move—the momentary effects of a football injury. Put in that perspective, just to be walking through the icefall on Mount Everest was a blessing for me. It made me think of how many people have given up on life or activities for whatever reason. It made me thankful that such a life-changing injury still did not keep me from moving forward. I feel blessed to have had this experience. This was a pinnacle of life for me.

Phil Henderson makes his way down through the Khumbu Icefall.

## ALWAYS WITH ME

There is not a day that goes by that I don't think about this expedition. For whatever reason, unlike my thoughts about any other expedition, I don't see it changing anytime soon. This was another life-changing experience for me. It will help me keep life in perspective. It is the best example I can think of to show why it is important for me to have expeditions as part of my life.

—PHIL HENDERSON A veteran mountaineer and senior field instructor for the National Outdoor Leadership School (NOLS), Phil Henderson was a member of the 50th Anniversary Legacy Climb team.

and the Lhotse Face have been more dangerous than usual. Several Sherpas have been seriously injured by falling rock on the Lhotse Face. One was hit by a collapsing serac, and the car-size chunk of ice broke an arm and leg. Two teams, including Himex, the largest on the mountain, have abandoned their attempts due to the excessive risks.

Snowstorms in early May finally, if temporarily, glue the mountain back together, and the Sherpas manage to fix lines up to the 25,900-foot South Col camp. On May 18, lines are strung up to the top by the Sherpas, led by none other than our own Kagi. The next day more than 200 people swarm to the summit. Four clients will die in the effort.

We hold our fire, watching, waiting, and recuperating at Base Camp. The next window of opportunity, according to satellite-assisted weather predictions, appears to be May 25. We rest and sleep and eat as much as possible. We are now reduced to a team of six, for Phil Henderson has developed a respiratory infection and magnanimously taken on the role of Base Camp manager.

Our summit push begins at the crack of night, 2 a.m., on May 21. We reach Camp II before noon and spend two days there—again eating and sleeping—before

**MARK JENKINS PASSES**
by debris left on Everest after
being beaten by winds above
Camp III on the Lhotse Face.

five of us ascend to Camp III. Conrad intends to come up the following day with the Sherpas.

## ASKING WHY

We spend hours cutting out an icy ledge in the sky. Camp I and Camp III are the only camps the Sherpas don't build for us. From our perch 23,000 feet above sea level, we peer down into the blinding whiteness of the Western Cwm. We spy the miniature tents of Camp II in the moraine, then out farther the crevasses around Camp I, and still farther to Base Camp, obscured by roiling clouds that resemble a slow-motion avalanche. The horizon, fading with great distance as in a watercolor painting, is a series of jagged ridgelines. We can see so far we can almost see into ourselves. A view you only get by getting up here.

After two months on Mount Everest, we are friends. We know each other's battles and betrayals—some of them anyway—and we've got each other's backs: When Hilaree sprained her ankle, when Sam lost heart, when I was too sick to sit up. This is what happens when you sleep beside each other in tiny tents on a long expedition. It's what is supposed to happen. You start out as individuals and transform each other into a team.

Camp III, halfway up the Lhotse Face, is like a skyscraper with no walls. No one else is on our floor, we have our own slice of solitude, although there are tents and climbers below us and above us.

Despite holing out directly beneath a serac that could collapse and crush us at any moment, this aerie is magical. There is not a breath of wind, which inexplicably has been the case every time we've been up here. The evening sun has turned the glacier into a wide, brilliant passageway, the flanks of Everest into the shimmering walls of a castle. The view is to die for, and many have.

"You know, at home, I ask myself why," says Hilaree, sitting on her haunches on the brink of emptiness, sipping coffee, her crampon points sunk into the snow. "Why climb mountains? Why leave my kids and my husband? Take all the risks . . ." An athlete on a dozen expeditions, Hilaree pauses, then sweeps an arm before her. "Well, this is why."

Kris, camera squeezed to his eye as usual, is scrambling back and forth on our slippery ledge.

"Right now, right here," he exclaims, "this is as good as it gets." With legs like an Olympic skier, he has survived six 8,000-meter ordeals.

Skinny Sam steps too close to the edge, and Emily, sitting cross-legged in the tent, shoving her blond hair under a wool cap, shouts at him. "Sam!" These two are the young guns, and neither has climbed above 13,000 feet before this expedition.

The landscape is glowing. We all are transfixed by the sublime view. There is a brutal, black-and-white beauty to snow and stone and nothing more. It is symbolic of the consequential simplicity of climbing mountains. Make the right decision, you live; make the wrong decision, you die. Most other sports are mere games. Mountaineering is mortal. There is luck but little forgiveness. Everything is distilled down to the elemental. Hot drinks, cold fingers,

**BUNDLED-UP MARK JENKINS** (above) takes a few minutes to chill outside his tent at Camp III on the Lhotse Face.

**A LINE OF** climbers stretches up the Lhotse Face (below). They approach the Yellow Band, a strip of limestone that angles up Everest at 20 to 30 degrees. Bottlenecks can occur at a number of places during the climb, including the Yellow Band, since teams traverse the same fixed path and there are only short climbing windows.

and Hilaree O'Neill rest at Camp III after a long day. High on the mountain, the team played a waiting game, anticipating the window when they could attempt the summit. That window came on May 23, 2012.

burning light, stench of sweat, rumble of avalanches. There is nothing alive at this altitude—the color green doesn't even exist—nothing alive but our determination to make it to the top.

## CAMP LIFE

The moment the sun sinks, the temperature plunges, but we're already in our tents inside our foot-deep sleeping bags. Two women and three men on the edge of the world. Sam and Emily and I are lying side by side. Sam is melting snow on the camp stove, Em is wagging her head to José González crooning through her earbuds, I'm writing. Next door, Kris and Hilaree are cutting chunks of cheese and prosciutto while singing along to the Raconteurs ripping from matchbox-size speakers. All the activities are interchangeable, as they have been whenever we have been at Camp III or Camp I. Soon I'll be melting ice and Em will be opening our food packets and Sam listening to Glitch Mob, Kris calling Base Camp for a weather report and Hils collecting hunks of ice for the pot and humming Johnny Cash to herself.

As darkness grips us like a cold fist, we are all hyperaware that in 24 hours—after 55 days of climbing—we will be going for the summit.

We've marched up and down the mountain four times to get our bodies in condition, but anything could still happen. One of us could get sick, or slip and fall, or simply become too exhausted to carry on. A rope could pull loose or a serac topple. It could snow a foot or a ferocious wind could turn us around. This is Everest, after all.

The only one thing we can control is our minds. Climbing Everest takes so long, just keeping your head in the game is half the battle. We switch off our headlamps and burrow down into our bags. Tomorrow it's finally showtime.

## DODGING THE CROWD

The next morning we awake to enthusiastic shouts. Our Sherpas have arrived from Camp II, ready to climb with us up to Camp IV. Donning our down suits and

### THEN & NOW ICE AX

| 1963 | 2012 |
|------|------|

The difference between the ice ax used in the 1963 Everest expedition and the sleek Black Diamond model used during the 2012 climb can be summed up in one word: design. Today's tool is lightweight aluminum with an ergonomic stainless-steel head, an aggressive pick, and a large hole for a carabiner. The older ax, with a wooden handle and steel head, weighs more.

t was pretty late in the game when I began telling friends and family that I was going to spend ten weeks of my spring climbing Mount Everest. I was met with many a blank stare and uncomprehending blinks. This reaction was not related to the fact that I wanted to climb Everest—anyone who knows me understood that attempting to climb to the highest point on Earth was inevitable for me. The expressions I was seeing had more to do with "how" versus "why."

The expedition only finalized for me in mid-January 2012. With an impending departure date of late March, how was I going to organize my life, let alone train for such a massive undertaking, in such a way that I could be away from home for ten weeks?

### LEAVING THE CHILDREN

This expedition came together at a rather inconvenient time for me. I am the mother of two small boys: In the winter of 2011–12, they were two and four years old. While I had always had the Khumbu and Everest on my mind, I'd envisioned it to be a few more years down the road, when my children were in school. No such luck. Even my husband agreed that this opportunity to climb Everest was too good to pass up. The team was outstanding, the sponsorship very solid, and the partners—The North Face, National Geographic, and the Mayo Clinic—were all too intriguing.

Together, my husband and I decided I should go. The next two months were absolute and utter chaos. Apart from the logistics of arranging child care and a support system for my husband while I was away, I needed to train. As a professional ski mountaineer I was not starting from scratch, but as a full-time mom I had a limited training window at best.

### TRIPLE CHALLENGE

Everest was not my first 8,000-meter peak. I had already climbed and skied Cho Oyu and attempted Gasherbrum II in Pakistan. From these experiences, I knew training for me was a threefold endeavor: physical fitness, mental toughness, and weight management.

O'Neill melts ice for drinking water.

In hindsight, the abominable snow conditions of the 2011–12 winter season actually worked in my favor. As a skier, if there is good skiing, that is the only activity I will do. But as a skier, I know that it gets me in shape for nothing other than skiing. The poor snow conditions allowed me to think outside the box when it came to getting in shape for Mount Everest. First and foremost, I needed a way to train while being with my two kids. At least three days a week I would take both boys skiing by myself. While my four-year-old could mostly manage on his own, my youngest was often between my legs or on a harness with me controlling his speed from behind. Both of these methods require serious snowplowing on my part, which is way more challenging than just skiing! The rest of the time, I would carry him.

For me, training is not only about physical fitness but also mental fitness. Carrying an often screaming and kicking 30-pound two-year-old around the mountain for hours at a time was torturous but amazing training physically and, especially, mentally.

As the winter shaped up a bit, I did go back to my normal standard of training: ski touring. This was pretty important to me as a tool for Everest because it was my intention to try to ski from the summit. With that in the back of my mind, I knew I needed to get on terrain that provided mental training in the form of exposure, and lots of it.

### THE TOLL ON THE BODY

One of the noted physiological responses to extreme altitude is a relentless loss of weight and overall deterioration of muscle. Having spent a lot of time at high altitude, I am very aware of the delicate balance between proper acclimation and the deterioration of the physical body. Many high-altitude climbers, including myself, try to give their bodies a bit of a cushion by putting on extra pounds before attempting an 8,000-meter peak. I call it the expedition diet. Eat anything and everything in hopes of prolonging your body from eating its own fat reserves and muscle. In my case, this is not so easy. One side effect of having children is that it changed my body type to be much leaner. It

O'Neill, high on Everest, makes her way to the summit.

is very difficult for me to put on weight, and my average weight is about five pounds less than it was pre-children. Therefore, eating was another important aspect of my preparation for Mount Everest.

The most challenging thing I find with balancing being a mom and trying to stay fit for big expeditions is that I am not in the dangerous, exposed terrain often enough to be as comfortable as I would like. With that said, I spent a fair amount of the time I had without my kids trying to ski very steep lines in the backcountry. I also found friends to take me ice climbing just to reacquaint myself with exposure and hanging from ropes and crampons and ice axes. I find

these types of training to be much more valuable to me than pure fitness.

In the end I felt relatively ready for the expedition. The final challenge was switching mental gears from mom to mountain climber—perhaps the hardest task of all.

—HILAREE O'NEILL Named by *Outside* magazine as one of the most adventurous women in sports, Hilaree O'Neill skis and climbs around the world as a North Face athlete.

crawling out of the tent, we are stunned to see an endless line of climbers passing near our camp. What the hell? Where did all these people come from? We pull on our harnesses and step in line, clipping our ascenders onto the string of ropes that rise up the Lhotse Face.

Below us I spot over a hundred climbers, trudging nose-to-butt like cattle. Above us struggle another hundred climbers, also in single file, dragging themselves up the ropes. We went to sleep on a mountain and woke up in a traffic jam!

I am wedged between an unknown climber above me and an unknown climber below me. Bumper to bumper, we all must move at exactly the same speed, regardless of strength or ability. I am appalled. This is a parody of mountaineering.

Of course, we knew that too many people were crowding the Everest route. That's why we waited for the second window of good weather—hoping most climbers would go for the top the first chance they got. Alas, hundreds of other climbers had the same idea.

Some problem has halted climbers on the line above me. The human highway stops. After ten minutes of standing still, I can take it no longer. I remove my ice ax, unclip from the lines, and swerve off on to open ice, roaring past one climber after the next. I overtake 10 climbers, 20 climbers. I am unroped, soloing, but the terrain is moderate. When I get to the end of the jam, I discover a Sherpa trying to help a fumbling climber who doesn't know how to unclip his ascender or reattach it to the next rope. I move past them and clip back onto the lines.

I soon reach the tail end of another clot of climbers. Once again I unclip and go around them. En route, I note that at least 30 people are hanging on a single ratty rope that is anchored by a single badly bent picket pounded into the ice. This is lunacy. If the picket popped, the rope or carabiner would instantly snap from the weight of three dozen falling climbers, and they would all cartwheel down the Lhotse Face to their deaths.

Separated from my own team, I continue to hopscotch up the ropes, crossing the steep, bare rock of the Yellow Band and traversing behind the Geneva Spur. I reach Camp IV, at 25,900 feet, before lunch. At least a hundred tents are spread out across the desolate South Col like some wind-blasted refugee camp. Debris is everywhere: garbage, tattered nylon flapping on the skeletons of abandoned tents, piles of frozen human excrement.

Our team's Sherpas have erected a pair of tents, and I collapse inside one of them. I take off my oxygen mask, lie back, and close my eyes. Kagi brings me a bowl of hot noodles. Panuru ducks his head into the tent and tells me to put my oxygen mask back on. One by one my teammates arrive, first Hilaree with Kris, both of whom crash inside the other tent. Emily shows up, then Conrad, lastly Sam. Conrad, the old lion, ascended all the way from Camp II to Camp IV without oxygen. He looks shattered, his face drawn and grizzled.

"I'm cooked," he croaks. "I can't go for the summit."

In a few hours we will set out, along with at least 200 others, and we know already it will be a mess. Above 26,000 feet, in the death zone, the safety margin is as thin as skin and rescue almost impossible. Our biggest fear is not for

## "IT IS HARD TO HAVE A FRIEND WHO IS NOT A CLIMBER. HE NEEDS TO HAVE SHARED THE MANY CLOSE-TO-DEATH EXPERIENCES, AND NOT HAVE PANICKED."

—DOUGAL HASTON, SCOTTISH MOUNTAINEER WHO, WITH DOUG SCOTT, MADE THE FIRST ASCENT OF THE SOUTHWEST FACE OF EVEREST

ourselves, but for all the climbers around us, too many of whom don't have the mountaineering experience or technical skills to be so high.

## THE PATH LESS TRAVELED

In 1963, Whittaker and Gombu had been utterly alone up high on the mountain; today this would be unimaginable. Relying heavily on Sherpa support and bottles upon bottles of supplemental oxygen, more than 300 people a year summit Everest, a total of almost 4,000 since 1963. (When all the numbers are compiled, more than 400 people will summit in the spring of 2012.) Sans oxygen—the blood doping of mountaineering—and Sherpas, Everest would still be the exclusive domain of the world's best high-altitude mountaineers. Like colorful Simone Moro, who, after forsaking the West Ridge and switching to the South Col, couldn't take the crowds and turned around. "If I want to stand in line, I can go to the grocery store," he would tell me later.

In contrast to the annual trail of bodies up Everest, the most respected ascents in modern alpinism are undertaken by a small, undeterrable, unsupported team that pushes the limits of skill and stamina. They lead all the pitches themselves, make all the decisions, take all the risks, and carry all their own accoutrements. "Everesting" is in many ways the antithesis of the best 21st-century mountaineering: Sherpas put up all the ropes, Sherpas put in all the camps, Sherpas carry all the loads, Sherpas do all the cooking, and Sherpas escort the climbers to the top. The Sherpas want the work and are well paid for it, but outsourcing so many of the hazards and so much of the heavy lifting has turned climbing Everest into a form of trophy hunting. It's like shooting an elephant. The elephant isn't going anywhere. It will just stand there. And with enough bush beaters and a big enough gun, you can kill it.

Luckily for skilled mountaineers, all guided climbers use only two standard routes on Everest, either the South Col in Nepal or the North Col in Tibet. These two routes, from an experiential point of view, are ugly and abused—littered, shit ridden, crowded. They are superhighways. Fixed ropes, put in by the Sherpas, run up both routes all the way to the summit, which is precisely why they are guided. Clients would not be able to climb Everest without them.

But don't get the wrong idea. Everest is hardly crawling with climbers. There are a dozen other routes on the mountain! Most of them were put up in the 1980s, and all of them are difficult and seldom climbed. Everest is a massive, majestic, wondrous peak with two tiny ant lines on it. You can still climb Everest and not see a soul—you must merely have the courage, capability, and experience to tackle a tough route.

# "WE NEED TO GET THE REAL ADVENTURES BACK."

-INES PAPERT, GERMAN ROCK AND ICE CLIMBER

**HILAREE O'NEILL (FRONT)** and Mark Jenkins take a breather after climbing the Yellow Band. O'Neill has shed her heavy coat: Temperatures can vary widely on Everest, from solar warmth to biting cold. Nuptse rises behind them.

# SUMMITING

At Camp IV, we rest in our tents until 9 p.m.: Then it's time. A frigid wind is roaring as we strap on our oxygen masks, and every patch of exposed flesh will be frostbitten. In our bulky down suits, goggles, respirators, and huge gloves, we resemble astronauts more than mountaineers.

As we set off in the swirling darkness, I look up at a Christmas tree of twinkling lights, climbers' headlamps rising straight up into the black sky. Reaching the fixed lines, we again find ourselves behind dozens of dangerously slow-moving clients.

Only an hour above Camp IV, we pass our first body. The dead person is lying on his side as if napping in the snow, his head covered by the hood of his parka. Ten minutes later we pass another body, her torso and head wrapped in a Canadian flag. Twenty minutes later, another corpse, a Korean. Still attached to the rope, he is sitting in the snow, frozen solid as stone, his face black, his eyes wide open.

Four people died on the first summit push a week ago. How? Why? In interviews with Sherpas, I will discover that in every case, the cause of death was arrogance. All four were told by their Sherpas that they were moving too slowly, that even if they reached the summit they would not have the strength to get back down—and all four refused to turn around.

The woman wrapped in the Canadian flag was said to have required nine bottles of oxygen (two or three is typical) before she collapsed. Tragically, pride goeth before the fall. Danuru told me that he has had clients in the past who, when told to turn around, said that they'd paid a lot of money for the trip and expected to reach the summit. When they finally collapsed, Danuru simply lassoed them and dragged them down through the snow. With the mountain so dry and rocky, this kind of rescue is now impossible. If you stop and sit, you freeze to death.

The Korean, when his Sherpa pleaded for him to turn around before he died, had actually punched the guide in the face. Now, stepping over his icy corpse, his face frozen in a rigid grimace, I overcome the gruesomeness of the scene by reminding myself that the mountain did not kill these climbers—they killed themselves. They died of hubris.

Panuru and I submit ourselves to the stop-and-go traffic for as long as we can stand it. Then Panuru taps me on the shoulder and motions for me to go around. Despite the fact that it is pitch-dark and we are on a knife-edge ridge of snow with 5,000-foot drops on either side, I gladly unclip and climb around a knot of ten climbers. Panuru follows, then passes me wordlessly, giving me the thumbs-up. From that point on, Panuru and I form a team. We use the ropes only when no one else is climbing directly above us. Otherwise, we solo around every bottleneck.

**CONRAD ANKER AND** Jangbu Sherpa (previous pages) retrieve gear that was stashed on the West Ridge for the summit attempt planned by Anker and Cory Richards. Poor conditions made the planned ascent of the West Ridge too dangerous, but Anker would summit from the South Col route without using supplemental oxygen.

**SAM ELIAS LOWERS** himself (opposite, at right) from the South Summit to continue on toward the top of the peak. At 28,707 feet (8,750 meters), the South Summit is the first place where Everest's true summit is visible. The climbers then head up the ridge, visible here behind Elias.

It is still dark, but the sky has turned purple when Panuru and I reach the South Summit. Now most of the headlamps are below us rather than above us. We traverse the col to the base of the Hillary Step, a 25-foot vertical wall of rock, scratched and scarred by thousands of crampon points. Several climbers are working their way up the lines. Panuru and I swiftly climb the rock and go around them.

We can see the summit now. It is just along the ridge crest. Pink velvety light is saturating the sky. I feel as if I'm in the center of a fish-eye lens, watching the world dropping away in all directions. Everything on Earth is below me. Everything. The rows and rows of mountains, even the clouds. The only thing above me is outer space.

Poking around a rock, we meet up with Kris, Hilaree, and Danuru. Jangbu has already gone up to take GPS measurements. The five of us walk slowly toward the summit together. The wind is tearing at the Tibetan prayer flags planted at the apex, but we can't get near them: At least 40 people have packed onto the summit hump. Instead we stop just a few feet below the top, throw our arms in the air, and tightly hug one another.

Emily, then Sam, as well as our other Sherpas—Dawa, Tendi, Mingma, Sonam Dorje, and Lakpa—will all stand on the summit in several hours. The next day, as if rising from the dead, Conrad will rally and summit without oxygen, a feat fewer than 200 climbers have accomplished. In a supreme test of endurance, Kris and Hilaree will push on to summit neighboring Lhotse after only a half day's rest at Camp IV. Our entire team, climbers and Sherpas, will rendezvous safely at Base Camp three days later.

At this moment, though, standing on the top of the world, I know I should feel elated, joyful. I know this is how the summiters of the '63 American expedition felt. "We hugged each other as tears welled up, ran down our oxygen masks, and turned to ice," wrote Hornbein. But climbing Everest was a bigger, richer challenge 50 years ago. There were more unknowns, the margins slimmer, the personal effort greater. The truth is, Everest has become domesticated, declawed. The majesty of the mountain and the meaning of climbing it have diminished. A half century ago, climbing Everest was a triumph of mountaineering prowess. Today, from the perspective of modern alpine ethics, it is almost an unsporting accomplishment.

So what do I feel standing on the summit of Everest? I admit to some pride for simply persevering, for sucking wind day after day and still putting one foot in front of the other. And I feel relief. The damn thing is done. But I also have an uneasy sense of ambiguity. I don't think I really deserve the grandeur of the world's highest peak. It wasn't a fair fight, and I know it.

**HILAREE O'NEILL PAUSES** (below) at the top of the world.

**MEMBERS OF THE** South Col Five walk along the corniced ridge of the South Summit (opposite) after their successful climb. Makalu Peak rises in the background at right, and Kanchenjunga can be seen in the distance. The team had a good day to summit, with low-lying clouds a sign of stable weather.

# CHAPTER 8

# THE FUTURE

David Breashears

# OF EVEREST

Everest, at left,
glows at sunset.

IN THE SPRING OF 2012 I SPENT A MONTH AT MOUNT EVEREST BASE CAMP FOR THE GLACIERWORKS PHOTOGRAPHIC PROJECT. I ARRIVED IN EARLY APRIL, JUST AS THE EXPEDITIONS WERE ESTABLISHING THE CAMPS.

In the spring of 2012 I spent a month at Mount Everest Base Camp for a Glacier-Works photographic project. I arrived in early April, just as the expeditions were establishing the camps for the climbers, clients, and guides who would soon follow. I wandered slowly up the rock-strewn glacier ice, pausing frequently to stare at the hundreds of brightly colored tents clustered tightly along the meandering trail. Most were small personal tents, but many were large dining, cooking, and communication tents. Generators purred in the distance, yak bells tinkled from the transport caravans, and a multitude of languages buzzed from the expedition camps. Thousands of prayers flags fluttered in the breeze. National flags from myriad nations rippled in the wind, too. It was an international village at the foot of the Khumbu Icefall. Hundreds of climbers would shortly be residing here, all with the same goal and each for a different reason. Yet Base Camp bore no resemblance to the one I visited 29 years earlier in the spring of 1983.

ICY PRAYER FLAGS wave in the breeze at Base Camp. As climbers and trekkers from all over the globe come and go, this international village is ever changing—and becoming increasingly populated.

AMERICANS PETER JAMIESON and Gerry Roach (at right) celebrate their 1983 ascent of Everest. They were part of filmmaker David Breashears's team when he transmitted the first live images from the peak. The scene at the summit was a different one from today: There were no visible mementos left behind by previous climbers, no jostling for space with larger climbing groups. "It appeared as if no one had been there before us," Breashears said.

# THE AGE OF INNOCENCE

That year we were a small American climbing team with an essential Sherpa support team. Our small encampment, nestled up under the icefall, was hard to find. We had no contact with the outside world other than the sporadic delivery of mail by runner from Lukla, 40 miles down the trail. We were the only team on the Nepali side of the peak. The mountain was ours and it was wonderful. We shared its slopes with only our teammates. The mountain and its challenges felt intimate and authentic, and at the same time so immense and grand.

During that expedition I reached the summit of Mount Everest for the first time on May 7, 1983. I had been assigned to the team by the ABC TV series *American Sportsman,* which sent me to the mountain to transmit the first live images from Earth's highest point. We set out from our high camp in the dawn's early hours. For some reason we chose to pioneer a shortcut directly up to the Southeast Ridge from the South Col, but we unexpectedly encountered thigh-deep snow and were slowed by the wearying efforts of wading through it. There were no fixed ropes above the South Col—we didn't need any. We only found 30 feet of wind-battered rope hanging down the Hillary Step.

My teammates—Ang Rita Sherpa, Peter Jamieson, and Gerry Roach—and I traded the exhausting duty of plowing a trail up through the soft, deep snow. Yet we all managed to remain close together and gained firmer snow at the South Summit at 3:30 p.m., where we readied for the final push. Larry Nielson had chosen to climb without bottled oxygen and was lagging behind. The traverse from the South Summit to the famed Hillary Step is long and exposed. Not until I put that intimidating final 40 feet of steep climbing behind me did I know we would succeed.

I well remember my last few steps. The peak was pristine, devoid of footprints, survey tripods, promotional banners, prayer flags, and mementos that would later clutter its summit. It appeared that no one had come before us. The wind was light, but an occasional gust reminded us of its potential power. We shared the piercing spectacle of Everest's harsh, implacable grandeur and a great solitude. Since arriving at Base Camp, we had taken more than 40 days to reach this moment; days of hardship, teamwork, self-reliance, and growing camaraderie. The five of us were but tiny dark dots barely visible from afar against the shimmering white peak. I gazed north out into the vast expanse of the Tibetan Plateau and thought of the early 1920s British expeditions. Their magnificent pioneering efforts to climb the mountain from Tibet had all ended in failure. Many lives were lost. Standing there, I knew my life had changed. I had attained a childhood aspiration. A long journey had ended. The world was at my feet. I was filled with a quiet awe and wonder, and a deep satisfaction, too.

## LONELY HEROES

After assembling the transmission equipment, I pointed the video camera at my companions as we stood astride the great peak. Then I captured Larry Nielson's labored final steps as he became the first American to reach the summit without supplementary oxygen. His was a noble effort, but it had taken a toll.

By now late afternoon, low-angle light slithered around the peaks below, casting long shadows over the valleys. The closest members of our support team were nearly 8,000 feet below in the advance base camp. Isolated? Yes. But also exhilarated with what we had just accomplished. Nor did we mind the enveloping darkness and heavy snow as we staggered down to tents on the South Col, exhausted but triumphant.

# "THEY WENT TO THE LIMITS OF THE HUMANLY POSSIBLE."

-DAG HAMMARSKJÖLD, UNITED NATIONS SECRETARY-GENERAL, *NATIONAL GEOGRAPHIC*, 1961

PETER JAMIESON WITH a frozen beard on the summit of Everest

I lay awake in my sleeping bag that night, recalling images of tired faces, exultant grins, and glinting peaks. Only hours earlier, I had become the 130th person to reach the top of Mount Everest in the 30 years since Edmund Hillary and Tenzing Norgay's 1953 ascent. I was proud to belong to that select club, but I thought it was already 100 members too big.

## THE END OF INNOCENCE

Since then I have returned to the mountain often and reached the summit four more times. I couldn't have imagined how much the Everest experience would change, and in so little time. Today nearly 129 people may reach the top of Everest in a single day; the total number of ascents recorded as of spring 2012 is more than 6,200.

Since 1856, when its supreme height of 29,002 feet (a number later revised to 29,035) was announced to the world, Mount Everest has exerted an enduring tug on the human imagination. I felt it too, when I was 11 years old, transfixed by a book's color photograph of Tenzing Norgay wearing the alien garb, and oxygen mask, of an explorer on the planet's outermost edge. In that single triumphant

CROWDS ON THE popular routes to the summit (below) are a continually growing safety concern. When good weather opens up, the routes to Everest are quickly jammed.

LESS EXPERIENCED CLIMBERS (opposite) work on their mountaineering techniques on the Khumbu Glacier. Larger guiding organizations set up "obstacle courses" for their clients to gain confidence and skills before ascending.

image I found an example that represented my own yearning for independence, discovery, and freedom. I sensed that the man behind the oxygen mask and thick down suit had taken risks and striven mightily to reach that spot, and that he was standing right where he wanted to be. Someday I wanted to stand there, too.

I was in excellent company. Countless people before and after me have projected their own values, dreams, and ambitions on the mountain. British explorers George Mallory and Andrew Irvine were pioneering a route up Everest's North Face when they died in 1924, just shy of the summit, leaving a mystery for the ages. When Tenzing and Hillary made the first ascent in 1953, the world was riveted. Willi Unsoeld and Tom Hornbein, in 1963, completed the first traverse of the Himalayan giant, one of the greatest achievements in mountaineering history. Reinhold Messner and Peter Habeler were the first to reach the summit without supplemental oxygen in 1978; Messner did so again, solo, unsupported, two years later. A Polish team made the first winter ascent in February 1980.

Everest has equally been a magnet for unusual characters with unlikely aims. The "mad Yorkshireman" Maurice Wilson, an eccentric World War I veteran with no climbing experience, flew illegally to India in 1934, intending to crash-land his DH Gipsy Moth into the glacier at the base of Everest and then climb to the top.

While the future of mountaineering on Everest remains unclear with global climate change transforming the high Himalaya, one can rest assured that Everest will, as it has done in the past, continue to captivate a world audience.

Fleshing out the richness of the Everest experience requires a look at the social environment of the mountain and recognizing that, as they have in the past, Sherpas will be decisive factors in any future effort. Most notably, Sherpas, the formerly agrarian inhabitants of Everest's southern foothills, are no longer simply load bearers on Everest's majestic features. In the last decade of increased commercial guiding on Everest, we have seen Sherpa mountaineers take on a decisively important and urgent role in the safe management of foreign mountaineers on expeditions.

Pete's daughter Cleo with Yangin Sherpa

Through on-the-job training and such organizations as the Nepal Mountaineering Association and the Khumbu Climbing Center, we are witnessing a transformation of knowledge and endowing of professional guiding skills to enthusiastic and capable climbers of Sherpa and other Nepali ethnicities. Sherpas have from the turn of the 20th century exhibited their remarkably adaptable physiology for climbing in the high Himalaya. Now, they will dominate the arena with superior climbing, guiding and organizational skills. This transformation will not take place overnight, but improvements have been noted most recently on expeditions where Sherpa guides do all the technical engineering of the climbing route, a task once relegated to the superiorly trained Western climbers leading the expeditions.

There is also the urgent objective of sustainability and the promotion of the mountain as an international symbol of environmental sensitivity and best practices in a wilderness setting. The Sherpa people will be driving the changes here as well, with great support from the visitors from the developed world—but the local population need to assume the role of champion stewards. Managing Base Camp habitation, organizing removal of waste materials, and ensuring the acceptable condition of the physical environment are critical objectives. Yet more expansively, conditions leading to Everest Base Camp in the villages of the Sherpa heartland need also to be addressed. Deforestation, overdependence on slowly renewable fuel sources, overconstruction of certain municipalities to increase tourism, management of potable water, and a vast array of similar issues face any future that Everest anticipates.

Though unmentioned, the political and social challenges that the Everest region, its people, and its visiting mountaineers must confront are complex. With a central government far away in Kathmandu that is characteristically unaware of the issues faced by Everest locals, new authority and resources need to be both delegated to and managed by the Sherpas themselves or committees of Nepali people intimate with the environmental and social challenges that threaten the communities, their livelihood, and the fabric of their fascinating society. Though the problems are large and solutions obscure, the Sherpa people have demonstrated, from one generation to the next, a tenuous adaptability and an enthusiastic and entrepreneurial spirit, tempered with deep compassion and strength. They are amongst the world's great survivors.

There's no telling what future awaits; suffice it to say that Everest has the potential once again to be not only the sine qua non of the mountaineering endeavor but also an example of enlightened development of rural communities and stewardship of one of the grandest mountains on Earth. Expect to be surprised, inspired, frustrated, perhaps incredulous—just not bored!

—**PETER ATHANS** Nicknamed "Mr. Everest," Pete Athans is a mountaineering athlete with The North Face who has summited Everest seven times. He and his wife, Liesl Clark, and their two children live part of every year in Khumbu.

But British authorities impounded his plane and denied him permission. Unde-terred, he trekked through Tibet to Everest with only a small support team. His body and tent were found in 1935 at 21,400 feet, just below the North Col.

Philosophy professor Woodrow Wilson Sayre, grandson of the President, recruited a geology student, schoolteacher, and lawyer to make a wildly amateur and ambitious attempt to climb Everest from the north in 1962. They entered Tibet illegally from Nepal, unsupported, ferrying all their own gear and food to the mountain's north side. Their audacious and little-known attempt reached nearly 25,500 feet before turning back. The exhausted team retraced their steps to Nepal, ending the first American expedition to Mount Everest.

On September 26, 1988, French mountaineer Jean-Marc Boivin completed the first descent from the summit by paraglider. Slovenian climber and extreme skier Davo Karničar climbed to the summit October 7, 2000, then descended on skis during the snowy post-monsoon climbing season. And the first marriage vows were pledged on summit by Pem Dorje Sherpa and Moni Mulepati on May 30, 2005.

Whatever one's motivation, reaching the top of the world has always been an ironman achievement. One must overcome the mountain's barriers, the

## THEN & NOW CAMERAS

### 1924

### 2012

Alexander Kellas, an early explorer of the Himalaya, used this vest-pocket camera on his unfulfilled Everest expedition of 1921; he died en route to the mountains of dysentery and heart failure. The Canon 5D Mark II camera with numerous digital features, including video, was one of several cameras used on the 2012 National Geographic expedition.

IN 1983, FILMMAKER David Breashears was the 136th person to reach Everest's summit. Today, during peak season, that many people may make it to the top in one day. A good weather window prompts more than 200 climbers to ascend the ropes up the Lhotse Face (above, in May 2012).

challenges of ferocious winds, thin air, and one's own physical limits and trepidations, to win the honor the public assigns to success on Earth's most exalted peak.

But in the 1980s those barriers began to fall. Two years after my first ascent, I returned to the mountain with Dick Bass, a strong high-altitude performer who after three attempts finally stood atop Everest, fulfilling his quest to reach the tallest summit on each of the seven continents. To the uninformed public, however, he was a Texas oilman who paid his way to the top. If Dick Bass can reach the summit, many began to reason, then so can I. In their eyes, the bar had been lowered. The numbers of climbers heading to the mountain each spring began to rise.

In 1996 I led the Everest IMAX film expedition and directed the filming on the mountain. I witnessed for the first time the early stages of overcrowding, along with widespread inexperience among the climbers and a strength-in-numbers approach. In this golden age of guided expeditions, 12 of them ventured up the mountain that season. In early May a violent, fast-moving storm left eight climbers dead, captivating the world's attention. The more the media and press dramatized what happened on Everest—a husband's last words to his wife, a doctor's loss of both hands to frost-bite, a perilous high-altitude helicopter rescue—the more compelling the mountain became, and its value rose in the eyes of the public, and climbers too.

Afterward, at every public lecture I gave, an audience member would ask, "Did that tragedy change the way people climb? The answer is, *Not really.*

What lies ahead for Everest? Every year the support system for climbers, clients, and guides will improve, along with better cooperation and communication among team leaders. Some guiding services will enhance their success rates, sharpen operational efficiencies, and increase their familiarity with their clients' strengths and weaknesses. But many will not.

Traditional and social media will continue to spread images and personal accounts of the Everest experience across the globe. In addition to the usual cadres of clients and guides, its slopes will be crowded with fund-raising, cleanup, and corporate-challenge expeditions but thinning prospects of achieving significant firsts.

Could a breaking point be reached? A photo taken on the Lhotse Face in May 2012 shows 156 people departing Camp III at 23,000 feet on their way to the South Col at 25,900 feet, streaming upward in a cheek-to-jowl procession. Can

TEAM MEMBERS OF the North Face/National Geographic 2012 climb head out together after acclimating at Camp II.

## "THE REAL GOAL IS TO BE SITTING NEAR THE HEARTH TOGETHER WITH THE LOCAL PEOPLE."

—CONRAD ANKER

the mountain and its popular routes sustain these numbers? Will climbers and clients begin to turn away from an experience that puts them in a conga-line formation? I know climbers who have reached the summit in the predawn darkness even though they had only a dim view from the summit. Why? They left the high camp at 8 p.m. to avoid the crowds.

I often think about my own role in the growing numbers of summit seekers. Some of us who treasure this mountain have spread our knowledge in the interest of income and education and inadvertently have become its prime publicists. Capt. John Noel, a member of the 1922 and 1924 Everest expeditions, made films of both journeys and crisscrossed the North American continent eight times to screen them for packed audiences. More than 70 years later, 30 million people in 38 countries would watch *Everest*, the IMAX film based on footage my team shot before and after the catastrophic storm of 1996.

Everest has always drawn the purist, the dreamer, and the trophy hunter. Most are a mix of all three, and each climber brings a wide range of abilities and capacity to the mountain. Dangers will increase if their ambition isn't matched by experience, and many individuals feel they don't need to take the necessary steps to prepare themselves for the mountain and its unforgiving moments. The public rarely inquires how you made it to the top, focusing only on the achievement. Too often climbers likewise focus on the prize and the accolades, not the huge support system and the Sherpas' hard work and expertise required to get them there. When climbers don't take part in the decision making, they are no longer agents in their own fate and forsake an intimate connection to the mountain, its weather, and its dangers.

# ENDURING REWARDS

In the spring of 2012 I talked with many climbers awaiting their first try for the summit. I met a single mother of three who explained that dreaming of the Everest adventure had got her through her toughest moments. Her children were grown and she was at Base Camp, trembling with excitement and full of apprehension. I also met a businessman who admitted matter-of-factly that he disliked everything about climbing, but was determined to get his made-it-to-the-top bragging rights.

I think about the enormous sacrifices these inexperienced climbers make, the expenses and time away from home and family. Yet this high-risk challenge demands preparation, teamwork, and training that enhances the rewards and value of the ascent. We become more acutely attuned to one another's actions and opinions, yet also independent and self-aware. In the end, however, every climber, whether heavily supported or not, must face the same cold and wind and make the same treacherous trips up and down the mountain. Everest is never an easy or comfortable experience. It requires fortitude to complete an unfamiliar challenge with all the inherent risks: thousands of feet of exposure, rockfall, ice collapse, avalanches, fast-moving storms.

Every year people die on the mountain. The summit attempts of spring 2012 left ten dead— some from inexperience or simple mistakes, some because their desire to reach the summit overrode internal circuits warning them to stop and turn around. From afar we judge and blame the victims for bad decisions and unpreparedness. But nobody goes there to die. Climbers get swept up in an all-consuming desire to reach the summit, willfully ignoring signals that say it's time to turn around. It's hard to make disciplined decisions in a hypoxic, sleep-deprived, and dehydrated state.

A YOUNG MONK at Tengboche monastery accepts filmmaker David Breashears's offer to take a wide-angle view through his camera (opposite). Breashears's IMAX film depicted the 1996 ascent of Everest by an international team that got caught in an infamous blizzard, the single deadliest event ever on the mountain.

FILMMAKING HAS BEEN a tradition on Everest since the earliest climbs. British Capt. J. B. Noel (below) photographed and filmed two expeditions in the 1920s, including George Mallory and Sandy Irvine's fateful attempt in 1924. He screened his films in North America, spawning early interest in the call to Everest.

**A CLIMB WITH** a view: The shadow of Everest aligns with the moon—a sight granted only to climbers who attempt the mountain. Despite the crowds, Everest is known for its great solitude.

**CLIMBERS APPROACH THE** Third Step, which lies at the base of the final pyramid, about 500 feet from the summit (following pages).

What are we seeking in that thin air? Is it an experience that we earn through craft, training, and perseverance? An escape from the dulling rituals of our lives down below? An extreme adventure? A transformative personal journey? Or just a feather in one's cap? Whatever one seeks, the ultimate goal, after all, is to return home safely. But while on that mountain, for those few glorious weeks, the climber experiences the simple rhythms of life, being and staying alive, the rising and setting of the sun, seeking food and shelter. We become more aware of changes in the weather and our bodies' needs and limitations. Our lives depend on it. A day's success is often attained by a simple repetitive act: Can you put one foot in front of the other? And after a few heaving gasps for air, can you do it again? You come away from this step-by-step struggle enriched by the surprises, the friendships forged along the way, and above all new knowledge of yourself and of our inherent resilience and fortitude.

## THE CALL OF EVEREST

Someone can certainly cast a negative image on the mountain based on today's concerns. But I will never see it like that. Perhaps the mountain will be overrun and overused. Or perhaps it will fall out of vogue, no longer a prize, eclipsed by other more prestigious and less common adventure challenges. In the end, I don't think it matters. Mount Everest can't be diminished. It can't be overrun. It isn't ours to control. We don't have the ability to change the shape of the mountain, but only to transform the idea of the mountain itself. The narrative of our experiences there is tightly woven into the fabric of our collective imagination. Will it remain so after the 20,000th ascent?

The monolithic ice-draped rock we call Everest is a towering screen onto which we project our hopes, dreams, and aspirations. But behind the screen is a cold, remote, disinterested peak that doesn't care if we scale it or not. It won't bend or break because of our presence or willfulness.

When I recall my first ascent in the fading afternoon light in 1983, the experience seems so innocent and private. Yet I can understand, and even anticipate, the deep satisfaction of the 10,000th person who reaches the summit, no matter what motivation takes him or her there. Intent is a private affair.

We human beings will forever yearn for transcendent experiences that fleetingly set us free from the complexity and confusion of everyday life. My hope is that the individuals who make the journey to this great mountain bring to it the experience, skills, and respect that it commands and deserves.

# "YOU NEVER CONQUER A MOUNTAIN ... YOU CONQUER YOURSELF."

–JAMES WHITTAKER, MEMBER OF THE 1963 AMERICAN MOUNT EVEREST EXPEDITION

## VOICES DAWN ON MOUNT EVEREST

One thing I hope will never change is the view on a clear day from the summit of Mount Everest just as the sun begins to rise. Change is inevitable, but there is something special about a sunrise on top of the world.

I can remember the frigid air, the type that takes your breath away—and I was already struggling to breathe. There was a persistent wind biting at the small bits of exposed flesh on my face, and the blue light of morning seemed to drag on forever. In the distance I could see a glow of color beginning to form on the horizon and then the tip of a triangular shadow slowly emerging across the view west. The sun's radiating strength warmed my smile as it crested the horizon, and I watched as the western expanse danced in the morning light. Slowly the shadow of the highest mountain in the world concealed everything in its path below. The triangle blanketing the mountains below, an indication that its presence is the highest, will always be unique to Mount Everest.

—**KRISTOFFER ERICKSON** A team member of the Legacy Climb, Kris Erickson is a photographer and explorer. After summiting Everest on May 25, 2012, he and teammate Hilaree O'Neill went on to summit neighboring Lhotse the very next day.

# FURTHER READING

Anker, Conrad, and David Roberts. *The Lost Explorer: Finding Mallory on Mount Everest.* Simon & Schuster, 1999.

Astill, Tony. *Mount Everest: The Reconnaissance 1935, the Forgotten Adventure.* Tony Astill, 2005.

Band, George. *Everest: 50 Years on Top of the World.* HarperCollins, 2003.

Bass, Dick. *Seven Summits.* Warner Books, 1986.

Bonington, Chris. *Everest Expeditions Omnibus.* London, Weidenfel and Nicolson, 2002.

Breashears, David. *High Exposure: An Enduring Passion for Everest and Unforgiving Places.* Simon & Schuster, 1999.

Breashears, David, and Audrey Salkeld. *Last Climb: The Legendary Everest Expeditions of George Mallory.* National Geographic, 1999.

Byers, A. C. "Contemporary Human Impacts on Alpine Ecosystems in the Sagarmatha (Mt. Everest) National Park, Khumbu, Nepal." *Annals of the Association of American Geographers* 95(1), pp. 112–140, 2005.

Coburn, Broughton. *Everest: Mountain Without Mercy.* National Geographic, 1997.

Coburn, Broughton. *The Vast Unknown: The First American Ascent of Everest.* Random House, 2013.

Coburn, Broughton. *Touching My Father's Soul: A Sherpa's Journey to the Top of Everest.* HarperSanFrancisco, 2000.

Davis, Wade. *Into the Silence: The Great War, Mallory, and the Conquest of Everest.* Knopf, 2011.

Fisher, James F. *Sherpas: Reflections on Change in Himalayan Nepal.* University of California Press, 1990.

Fürer-Haimendorf, C. von. *The Sherpas of Nepal: Buddhist Highlanders.* John Murray, 1964.

Gansser, A. *Geology of the Himalayas.* Interscience Publishers/John Wiley and Sons, 1964.

Gillman, Peter, ed. *Everest: The Best Writing and Pictures from Seventy Years of Human Endeavour.* Little, Brown and Company, 1993.

Gillman, Peter. *Everest: Eighty Years of Triumph and Tragedy.* The Mountaineers Books, 2001.

Hornbein, Thomas F. *Everest: The West Ridge.* 50th Anniversary Edition. The Mountaineers Books, 2013.

Hultgren, Herbert N. *High Altitude Medicine.* Hultgren Publications, 1997.

Isserman, Maurice, and Stewart Weaver. *Fallen Giants.* Yale University Press, 2008.

Jeffries, M. *The Story of Mount Everest National Park.* Cobb/Horwood Publications, 1984.

Keay, John. *The Great Arc: The Dramatic Tale of How India was Mapped and Everest was Named.* HarperCollins, 2000.

Klatzel, Frances. *Gaiety of Spirit: The Sherpas of Everest.* Rocky Mountain Books, 2010.

Krakauer, Jon. *Into Thin Air: A Personal Account of the Mt. Everest Disaster.* Villard Books, 1997.

Kukuczka, Jerzy. *My Vertical World.* The Mountaineers Books, 1992.

Lewis-Jones, Huw. *Mountain Heroes: Portraits of Adventure.* FalconGuides, 2011.

Lowe, George, and Huw Lewis-Jones. *The Conquest of Everest: Original Photographs from the Legendary First Ascent.* Thames & Hudson, 2013.

Macfarlane, A., R. B. Sorkhabi, and J. Quade, eds. *Himalaya and Tibet: Mountain Roots to Mountain Tops.* The Geological Society of America, Special Publication 328, 2009.

McDonald, Bernadette. *Freedom Climbers.* Rocky Mountain Books, 2011.

McDonald, Bernadette, *I'll Call You in Kathmandu.* The Mountaineers Books, 2005.

Rose, David, and Ed Douglas. *Regions of the Heart.* Michael Joseph, 1999.

Stevens, S. *Claiming the High Ground: Sherpas, Subsistence, and Environmental Change in the Highest Himalaya.* University of California Press, 1993.

Summers, Julie. *Fearless on Everest.* The Mountaineers Books, 2000.

Tenzing, Tashi. *Tenzing Norgay and the Sherpas of Everest.* Ragged Mountain Press, 2003.

Ullman, James Ramsey. *Man of Everest: The Autobiography of Tenzing.* Hazell Watson and Viney Ltd., 1955.

Unsworth, Walt. *Everest: A Mountaineering History.* Houghton Mifflin, 1981.

Webster, Ed. *Snow in the Kingdom: My Storm Years on Everest.* Mountain Imagery, 2000.

West, John B. *High Life: A History of High-Altitude Physiology and Medicine.* Oxford University Press, 1998.

Whittaker, Jim. *A Life on the Edge.* The Mountaineers Books, 1999.

# CONTRIBUTORS

CONRAD ANKER is one of America's best known alpinists. Born in San Francisco in 1962, he first climbed the granite domes and sheer faces of Yosemite Valley near Big Oak Flat, where his great-grandfather pioneered. His parents, outdoors enthusiasts, encouraged his passion for climbing. Anker graduated from the University of Utah while pursuing expedition climbing around the globe. In 1999, climbing Everest, Anker found the body of early British explorer George Mallory, who was last seen nearing the summit in 1924. He is co-author of *The Lost Explorer* and is a key character in the National Geographic film *The Wildest Dream: Conquest of Everest.* Conrad Anker is a professional athlete for The North Face and lives in Bozeman, Montana, with his wife, Jennifer Lowe-Anker, and their sons, Max, Sam, and Isaac.

THOMAS HORNBEIN, M.D., is an emeritus professor of medicine at the University of Washington. An anesthesiologist specializing in the physiology of breathing, he has always pursued mountaineering along with his professional career. In May 1963, he and Willi Unsoeld, members of the first American expedition to Everest, became the first climbers to ascend Mount Everest via the West Ridge. A span on that mountainside has since been named for him: the Hornbein Couloir. Hornbein wrote of his climbing experience in *Everest: The West Ridge,* issued in a third edition for the 50th anniversary of the expedition, in 2013. As he enters his eighth decade, Hornbein remains active in exploring, climbing, and caring for mountain environments.

DAVID BREASHEARS is a filmmaker, mountaineer, and executive director of GlacierWorks, a nonprofit concerned about climate change in the Himalaya. In 1983, he transmitted the first live television pictures from the summit of Everest; in 1985 he became the first American to summit Everest twice. He also co-directed and co-produced the first IMAX film shot on Mount Everest, which premiered in March 1998 and recounted the tragic 1996 climbing season. He is the author of *High Exposure: An Enduring Passion for Everest*, co-author of National Geographic's *Last Climb*, and contributor to National Geographic's *Everest: Mountain Without Mercy.*

ALTON C. BYERS, PH.D., a National Geographic explorer, is a mountain geographer, photographer, writer, and climber with more than 30 years of experience working in major mountain regions throughout the world. His areas of expertise include applied research, community-based alpine conservation and restoration methods, and climate-change impacts in high mountain environments. He has published and lectured widely on topics dealing with mountain conservation, culture, exploration, and climate change.

BROUGHTON COBURN has worked in development, environmental conservation, and protected area management in Nepal, Tibet, and India for 20 years and has overseen several charitable projects in Sagarmatha National Park. He has written or edited seven books, including National Geographic's best-selling *Everest: Mountain Without Mercy* and T*riumph on Everest,* a biography of Sir Edmund Hillary for children. He collaborated with Jamling Tenzing Norgay on the book T*ouching My Father's Soul: A Sherpa's Journey to the Top of Everest.* His book *The Vast Unknown: The First American Ascent of Everest,* on the 1963 American Mount Everest Expedition, will be published by Random House in 2013.

MARK JENKINS is a contributing writer for *National Geographic* magazine specializing in difficult or dangerous assignments. The author of four critically acclaimed books— *A Man's Life, The Hard Way, To Timbuktu,* and *Off the Map*— Jenkins has won numerous literary awards including the National Magazine Award, the Banff Literary Award, and the American Alpine Club Literary Award. He has covered the guerrilla war in the Congo, land mines in Cambodia, AIDS in Botswana, and koalas in Australia, and has undertaken more than 50 mountaineering expeditions around the world, from Tibet to Bolivia, Afghanistan to Uganda, and Greenland to Pakistan.

BRUCE JOHNSON, PH.D., is a professor of medicine and physiology at the Mayo Clinic and a consultant in the clinic's Division of Cardiovascular Diseases. He runs the Human Integrative and Environmental Physiology Laboratory and has a long history of studying limiting factors in human performance. He has set up mobile laboratories throughout the world, including in Antarctica, South America, and Nepal. He has published and lectured widely on human physiology related to health and disease.

DAVID R. LAGESON, PH.D., is a professor of geology at Montana State University. He received his doctorate from the University of Wyoming in 1980 and has taught and conducted research at Montana State for 32 years. His specialty is structural geology and tectonics. He is a fellow of the Geological Society of America and the Geological Society of London.

BERNADETTE MCDONALD, former vice president for mountain culture at the Banff Centre in Banff, Alberta, has authored eight books on mountaineering and mountain culture. Her numerous awards include the Boardman Tasker Prize and the Banff Mountain Book Festival Grand Prize, awarded for *Freedom Climbers* in 2011; Italy's ITAS Prize in 2010; and India's Kekoo Naoroji Award in 2008, 2009, and 2012. The American Alpine Club awarded her its highest literary honor for excellence in mountain literature. McDonald spends her discretionary time in the mountains, climbing, ski touring, and hiking.

2-3, Tommy Heinrich; 6, Edmund Hillary, Courtesy Royal Geographical Society (with IBG); 8, Barry Bishop/NGS; 9, Barry Bishop/NGS; 10, Barry Bishop/NGS; 11, Jim Detterline; 12-3, Barry Bishop/NGS; 15, C. Richards Photography; 16, J.B. Noel, Courtesy Royal Geographical Society (with IBG); 17, John Noel Photographic Collection; 18, Barry Bishop/NGS; 20, Aurora Photos/Alamy; 21 (LE), Barry Bishop/NGS; 21 (RT), Mark Jenkins; 22, Jimmy Chin; 23, Max Lowe; 24, © Eddie Bauer. Photo by Jake Norton; 25, Kristoffer Erickson; 26 (LE), Courtesy Royal Geographical Society (with IBG); 26 (RT), Summit Oxygen (www.summitoxygen. com); 27, C. Richards Photography; 28-9, William Thompson; 31 (UP), Christoph Hormann/Science Source; 31 (LO), Andy Bardon Photography; 32, Phil Henderson; 33, Jason Edwards/NGS; 34, James L. Amos/NGS; 36 (UP & LO), Courtesy Royal Geographical Society (with IBG); 37, Courtesy Royal Geographical Society (with IBG); 38, Courtesy Royal Geographical Society (with IBG); 39, Courtesy Royal Geographical Society (with IBG); 40, Courtesy Royal Geographical Society (with IBG); 41, Courtesy Royal Geographical Society (with IBG); 42, ETH-Bibliothek Zurich, Image Archive; 43 (UP), Courtesy Royal Geographical Society (with IBG); 43 (LO), Andy Ballingall; 45 (LE), Photograph by Fritz Müller, archives of Alton C. Byers; 45 (RT), Andy Bardon Photography; 46, Andy Bardon Photography; 48 (LE), Courtesy Royal Geographical Society (with IBG); 48 (RT), Courtesy Trimble (http:// www.trimble.com/); 49, Galen Rowell/ Corbis; 50, Kristoffer Erickson; 52-3, Robert Preston Photography/Alamy; 56-7, Robb Kendrick Photography; 59 (UP), Robb Kendrick Photography/NGS; 59 (LO), Robb Kendrick Photography; 60, Richard I'Anson/Lonely Planet Images/ Getty Images; 61, 1956 photograph by Fritz Müller, archives of Alton C. Byers; 64 (UP), Robb Kendrick/Aurora; 64 (LO), Photograph by Fritz Müller, archives of Alton C. Byers; 65 (LE), Courtesy Royal Geographical Society (with IBG); 65 (RT), Marco Simoni/Robert Harding/Newscom; 66, Max Lowe; 67, Max Lowe; 68, Robb Kendrick Photography/NGS; 70, Robb Kendrick/Aurora Photos; 71, Kristoffer Erickson; 72 (LE, CTR, & RT), Andy Bardon Photography; 73 (LE), Archives of Alton Byers; 73 (RT), Scott Warren/Aurora

Photos; 74, Andy Bardon Photography; 76, Peter McBride/Aurora Photos; 77 (UP), John Van Hasselt/Corbis; 77 (LO), Nawang Doka Sherpa; 78, Dieter Glogowski/laif/Redux Pictures; 79, Robb Kendrick Photography/NGS; 80, Robb Kendrick Photography; 82-3, Painting by Temba Sherpa, Kunde village, Sagarmatha (Everest) National Park, Nepal; 84, C. Richards Photography; 86, Andy Bardon Photography; 87 (UP), Tommy Heinrich; 87 (LO), Pasang Temba Sherpa; 88-9, Colin Monteath/Minden Pictures/NGS; 91, Steve Winter/NGS; 92, Ethan Welty/ TandemStock.com; 93, Alex Treadway/ National Geographic/Getty Images; 94, The Asahi Shimbun Premium Archive via Getty Images; 95 (UP), Anne B. Keiser/ NGS; 95 (LO), Jeff Davids/National Geographic My Shot; 96-7, Ethan Welty/ TandemStock.com; 98, Dan Rafla/Aurora Photos/Corbis; 99, NASA; 100, Mark Williamson/Getty Images; 101, Image by Robert Simmon, based on data © 2003 Geoeye. Courtesy of NASA; 102, Colin Monteath/Hedgehog House/Minden Pictures/Getty Images; 103, Steve Satushek/Getty Images; 104, Gordon Wiltsie/National Geographic/Getty Images; 105, Keren Su/China Span/Getty Images; 106, Blazej Lyjak/Shutterstock; 107, Alex Treadway/NGS; 108, 1956 photograph by Fritz Müller; archives of Alton C. Byers; 109 (UP), 1956 photograph by Fritz Müller; archives of Alton C. Byers; 109 (LO), Robb Kendrick/Aurora Photos; 110, Urban Golob/Anzenberger/Redux Pictures; 111 (UP), Kristoffer Erickson; 111 (LO), Martin Price; 112, Tim Watson/ National Geographic My Shot; 114, David Schweitzer/Getty Images; 115, Robb Kendrick/Aurora Photos; 116, Jimmy Chin; 117 (UP & LO), Jimmy Chin; 118, Paul Harris/AWL Images/Getty Images; 119 (UP), TAO Images Limited/Getty Images; 119 (LO), Eric PHAN-KIM/Getty Images; 120, Photograph by Alton C. Byers; 121 (UP & LO), Photograph by Alton C. Byers; 122, Photograph by Erwin Scheider, courtesy of the Association for Comparative Alpine Research, Munich, Germany; archives of Alton C. Byers; 123, Photograph by Alton C. Byers; 124, NASA; 125, Photograph by Fritz Müller, archives of Alton C. Byers; 126-7, Photograph by Fritz Müller, archives of Alton C. Byers; 127, Photograph by Fritz Müller, archives of Alton C. Byers; 128, Photograph by Fritz Müller, archives of

Alton C. Byers; 129, Photograph by Fritz Müller, archives of Alton C. Byers; 130, Michael S. Lewis/Corbis; 131 (UP), Urban Golob/Anzenberger/Redux Pictures; 131 (LO), Max Lowe; 132, Spcc/Xinhua/ ZUMAPRESS.com/Newscom; 133, Ethan Welty/TandemStock.com; 134, Kyohei Mitazaki/National Geographic My Shot; 135, Tommy Heinrich; 136-7, Jimmy Chin; 139 (UP), Luther G. Jerstad/NGS; 139 (LO), Mallory and Irvine Expedition/Jim Fagiolo/ Getty Images; 140 (LE), The Salkeld Collection; 140 (RT), Irvine Archive; 141, George Leigh Mallory, Courtesy Royal Geographical Society (with IBG); 142-3, The Alpine Club; 143, Mallory/Irvine Expedition/Jim Fagiolo/Liaison Agency; 146, T. Howard Somervell, Courtesy Royal Geographical Society (with IBG); 147, Mallory/Irvine Expedition/Jim Fagiolo/ Liaison Agency; 148, Noel.E. Odell, Courtesy Royal Geographical Society (with IBG); 149, Courtesy Royal Geographical Society (with IBG); 150, Courtesy Royal Geographical Society (with IBG); 151 (LE), Mark Theissen/Becky Hale, NGS; 151 (CTR), The North Face; 151 (RT), The North Face; 152 (LE), Mark Theissen/Becky Hale, NGS; 152 (RT), Courtesy Black Diamond Equipment; 153 (UP), National Portrait Gallery Picture Library, London; 153 (LO), Paula Paisley/The Cumberland News and Star; 154, The New York Times/Redux Pictures; 156-7, Courtesy Royal Geographical Society (with IBG); 158, HO/ Reuters/Corbis; 159 (UP), Michael S. Lewis/Corbis; 159 (LO), Patrick Riviere/ Getty Images; 160 (UP), Barry Bishop/NGS; 160 (LO), Barry Bishop/NGS; 161, Barry Bishop/NGS; 162-3, Barry Bishop/NGS; 164, Barry Bishop/NGS; 165 (UP), Nawang Gombu/NGS; 165 (LO), William F. Unsoeld/NGS; 166, Tony Riley/Chris Bonington Picture Library; 167, Tony Riley/ Chris Bonington Picture Library; 168, AP Images; 169 (UP), Photo by Jon Krakauer, taken on Royal Flush near Frisco, CO on August 3, this year; 169 (LO), Dale Johnson; 170 (LE), Mark Theissen/Becky Hale, NGS; 170 (RT), Claudio Baldini/Shutterstock; 171 (UP), Bogdan Jankowski; 171 (LO), Lars Baron/Bongarts/Getty Images; 172 (UP), Vincent J. Musi/NGS; 172 (LO), Nena Holguin, NGS; 174, Simone Moro; 175, Data Source: NOAA, Processing Meteotest; 176, Jimmy Chin; 177 (UP), Andy Bardon Photography; 177 (LO), Ryan Hill; 179, Stephen Venables; 180, Ed Webster/Mtn.

Imagery; 181, David Breashears; 182 (UP & LO), Jimmy Chin; 183, Jean Troillet; 184, Robert Schauer; 185, Neil Beidleman/ Woodfin Camp & Associates; 186, Leo Dickinson; 187 (UP), Max Lowe; 187 (LO), UK History/Alamy; 188 (UP), © COLLECTION MARCO SIFFREDI; 188 (LO), Didrik Johnck/Corbis; 189, Babu and Lakpa Sherpa; 190 (LE), Mark Theissen/Becky Hale, NGS; 190 (RT), The North Face; 191, Jimmy Chin; 192, Picture Norgay Archive/ Reuters; 193, Tim Stelzer/Getty Images; 194-5, Andy Bardon Photography; 197 (UP), Andy Bardon Photography; 197 (LO), Ted Wood; 198, Copyright 2012 Mayo Foundation for Medical Education and Research; used with permission; 199, Courtesy Royal Geographical Society (with IBG); 200, C. Richards Photography; 201, Stephen Venables; 202, Andy Bardon Photography; 203 (LE & RT), Mark Theissen/Becky Hale, NGS; 204, Ed Webster/Mtn. Imagery; 205 (UP), Copyright 2012 Mayo Foundation for Medical Education and Research; used with permission; 206, Bartosz Hadyniak/iStockphoto; 207 (UP), Ed Webster/Mtn. Imagery; 207 (LO), Mike Banks; 208, Design Pics/Sean White/Getty Images; 209, Jim Richardson/NGS; 210, Kristoffer Erickson; 211 (UP), Stephen Venables; 211 (LO), Andy Bardon Photography; 212, Barry Bishop/NGS; 213, Copyright 2012 Mayo Foundation for Medical Education and Research; used with permission; 214, Andy Bardon Photography; 215 (UP), Andy Bardon Photography; 215 (LOLE), Mark Theissen/Becky Hale, NGS; 215 (LORT), The North Face; 216-7, Andy Bardon Photography; 218, Jake Norton/ MountainWorld Photography/Aurora Photos; 219, Craig Kassover/National Geographic My Shot; 220, Andy Bardon Photography; 221, Andy Bardon Photography; 222, David L. Dingman/NGS; 223, David L. Dingman/NGS; 224-5, C. Richards Photography; 227, C. Richards Photography; 228, Andy Bardon Photography; 229, Andy Bardon Photography; 230, Barry Bishop/NGS; 231, Daniel E. Doody/NGS; 232, Barry Bishop/ NGS; 233, Barry Bishop/NGS; 234, George F. Mobley/NGS; 235 (UP), Barry Bishop/ NGS; 235 (LO), Grayson Schaffer; 236 (UP), C. Richards Photography; 236 (LO), Andy Bardon Photography; 237, C. Richards Photography; 238 (UP), Andy Bardon Photography; 238 (LO), Andy Bardon

Photography; 240, Kristoffer Erickson; 241, C. Richards Photography; 243 (UP), Emily Harrington; 243 (LO), C. Richards Photography; 244, Andy Bardon Photography; 245, Andy Bardon Photography; 246-7, Andy Bardon Photography; 248, Kristoffer Erickson; 249 (LE), Courtesy Royal Geographical Society (with IBG); 249 (RT), C. Richards Photography; 250, Andy Bardon Photography; 251 (UP), Kristoffer Erickson; 251 (LO), C. Richards Photography; 252 (LE & RT), Andy Bardon Photography; 253 (LE & RT), Andy Bardon Photography; 254, Kristoffer Erickson; 255, Kristoffer Erickson; 256 (UP), Phil Henderson; 256 (LO), Max Lowe; 257, Kristoffer Erickson; 258, Kristoffer Erickson; 259 (UP & LO), Kristoffer Erickson; 260, Kristoffer Erickson; 261 (LE), Mark Theissen/Becky Hale, NGS; 261 (RT), Courtesy Black Diamond Equipment; 262, Kristoffer Erickson; 263 (UP), Kristoffer Erickson; 263 (LO), Adam Clark; 264, Ralf Dujmovits; 266, Kristoffer Erickson; 268-9, Andy Bardon Photography; 270, Kristoffer Erickson; 272, Kristoffer Erickson; 273, Kristoffer Erickson; 274-5, Chris Noble/Getty Images; 277, Mark Jenkins; 278, David Breashears; 279, David Breashears; 280, Nawang Sherpa/Bogati/ ZUMA/Corbis; 281, Kristoffer Erickson; 282 (UP), Kristoffer Erickson; 282 (LO), Kristoffer Erickson; 283 (LE), Courtesy Royal Geographical Society (with IBG); 283 (RT), Matt Propert, NGS; 284, Andy Bardon Photography; 285, Kristoffer Erickson;  286, David Breashears; 287, Courtesy Royal Geographical Society (with IBG); 288, Tommy Heinrich; 289, C. Richards Photography; 290-1, Harry Kikstra/ExposedPlanet.com.

## Map Acknowledgments

**Front Matter** pp. 4-5, The Route Up Everest, *National Geographic Magazine*, Heinrich C. Berann, October 1963, p.470-471; **Chapter 1** p. 19, The Highest Peak, *National Geographic Everest*, 1997; **Chapter 2** p. 35, 470 Million Years Ago (Middle Ordovician), Ron Blakey; p. 44, Plate Tectonics Fault Boundaries, Dèzes, Pierre. *Tectonic and Metamorphic Evolution of the Central Himalayan Domain in Southeast Zanskar (Kashmir, India)*. Lausanne, Suisse: Section Des Sciences De La Terre, Université De Lausanne, 1999. Online: http://comp1.geol.unibas. ch/~zanskar/CHAPITRE2/page21.html.; Taylor, M., and A. Yin." Active Structures of the Himalayan-Tibetan Orogen and Their

Relationships to Earthquake Distribution, Contemporary Strain Field, and Cenozoic Volcanism." *Geosphere* 5.3 (2009): 199-214. Online: http://www.geo.ku.edu/programs/ tectonics/taylor/pubs/GEOS_00217. pdf.; Consultant: Micah J. Jessup, PhD, Assistant Professor, Department of Earth and Planetary Sciences, University of Tennessee, Knoxville, TN, 37996-1410 USA; p. 54, Eurasian/Indian Plate, Dèzes, Pierre. *Tectonic and Metamorphic Evolution of the Central Himalayan Domain in Southeast Zanskar (Kashmir, India)*. Lausanne, Suisse: Section Des Sciences De La Terre, Université De Lausanne, 1999. Online: http://comp1.geol.unibas.ch/~zanskar/ CHAPITRE2/page21.html.; Molnar, P., and P. Tapponnier." Cenozoic Tectonics of Asia: Effects of a Continental Collision: Features of Recent Continental Tectonics in Asia Can Be Interpreted as Results of the India-Eurasia Collision." Science 189.4201 (1975): 419-26. Print.; Consultant: Micah J. Jessup, PhD, Assistant Professor, Department of Earth and Planetary Sciences, University of Tennessee, Knoxville, TN, 37996-1410 USA; **Chapter 3** pp. 62-63, 1963 Expedition, *National Geographic Magazine*, October 1963, p.462-463; Base Map Sources: http://geoportal.icimod.org/Downloads/, http://www.arcgis.com/home/item.htm l?id=dd56c4ddea0b473bbbd188fdd7b 0ad72; p. 81, Himalayan Trust Project, *National Geographic Magazine*, May 2003, p.56; **Chapter 4** p. 113, Sagarmatha National Park, *National Geographic Magazine*, June 1982, p.699; **Chapter 5** p. 144-145, 15 Ways to the Top, *National Geographic Magazine*, May 2003, p.14-15, p. 155, Tenzing and Hillary's Route: The 1953 British Expedition up Everest, *National Geographic Magazine*, July 1954, p.12; p. 173, Reinhold Messner's First Solo Ascent, *National Geographic Magazine*, October 1981, p.557; **Chapter 6** p. 205, Life and Death on Everest, *National Geographic Magazine*, May 2003, p. 26; http://www.cbc.ca/news/interactives/ everest/ (source data: Jurgalski | 8000ers. com); 2011 and 2012 fatality data: http:// en.wikipedia.org/wiki/List_of_deaths_on_ eight-thousanders; http://www.alanar-nette.com/blog/; 2012 ascent data: http:// www.asian-trekking.com/about-us/more-about-asian-trekking/ang-tshring-writes/ item/296-spring-2012-expedition-facts-statistics-economic-contributions-note-worthy-climbs.html

**The Call of Everest**
Conrad Anker

### Published by the National Geographic Society

John M. Fahey, Chairman of the Board and
    Chief Executive Officer
Timothy T. Kelly, President
Declan Moore, Executive Vice President;
    President, Publishing and Travel
Melina Gerosa Bellows, Executive Vice President; Chief
    Creative Officer, Books, Kids, and Family

### Prepared by the Book Division

Hector Sierra, Senior Vice President and General
    Manager
Janet Goldstein, Senior Vice President and Editorial
    Director
Jonathan Halling, Design Director, Books and Children's
    Publishing
Marianne R. Koszorus, Design Director, Books
Susan Tyler Hitchcock, Senior Editor
R. Gary Colbert, Production Director
Jennifer A. Thornton, Director of Managing Editorial
Susan S. Blair, Director of Photography
Meredith C. Wilcox, Director, Administration and Rights
    Clearance

### Staff for This Book

John Paine, Text Editor
Elisa Gibson, Art Director
Matt Propert, Illustrations Editor
Carl Mehler, Director of Maps
Michael McNey and Jonathan K. Nelson, Map Research
    and Map Production
Julie Beer and Michelle Harris, Picture Legends Writers
Marshall Kiker, Associate Managing Editor
Judith Klein, Production Editor
Lisa A. Walker, Production Manager
Galen Young, Illustrations Specialist
Katie Olsen, Production Design Assistant
Margaret Krauss, Editorial Assistant

### Manufacturing and Quality Management

Phillip L. Schlosser, Senior Vice President
Chris Brown, Vice President, NG Book Manufacturing
George Bounelis, Vice President, Production Services
Nicole Elliott, Manager
Rachel Faulise, Manager
Robert L. Barr, Manager

The National Geographic Society is one of the world's largest nonprofit scientific and educational organizations. Founded in 1888 to "increase and diffuse geographic knowledge," the Society's mission is to inspire people to care about the planet. It reaches more than 400 million people worldwide each month through its official journal, *National Geographic,* and other magazines; National Geographic Channel; television documentaries; music; radio; films; books; DVDs; maps; exhibitions; live events; school publishing programs; interactive media; and merchandise. National Geographic has funded more than 10,000 scientific research, conservation and exploration projects and supports an education program promoting geographic literacy. For more information, visit www.nationalgeographic.com.

For more information, please call 1-800-NGS LINE (647-5463) or write to the following address:

National Geographic Society
1145 17th Street N.W.
Washington, D.C. 20036-4688 U.S.A.

For information about special discounts for bulk purchases, please contact National Geographic Books Special Sales: ngspecsales@ngs.org

For rights or permissions inquiries, please contact National Geographic Books Subsidiary Rights: ngbookrights@ngs.org

Library of Congress Cataloging-in-Publication Data

Anker, Conrad.
  The call of Everest : the history, science, and future of the world's tallest peak / Conrad Anker ; foreword by Thomas Hornbein.
    pages cm
  Includes bibliographical references and index.
  ISBN 978-1-4262-1016-7
  1.  Everest, Mount (China and Nepal) 2.  Mountaineering expeditions--Everest, Mount (China and Nepal)--History. 3. Everest, Mount (China and Nepal)--Description and travel.  I. Hornbein, Thomas. II. Title.
  DS495.8.E9A64 2012
  915.496--dc23

                                2012045336

Printed in the United States of America

13/RRDW-LPH/1